T-34

T-34

An Illustrated History of Stalin's Greatest Tank

Wolfgang Fleischer

Foreword by
Anthony Tucker-Jones

Greenhill Books

*T-34: An Illustrated History of
Stalin's Greatest Tank*
This English-language edition
first published in 2020 by
Greenhill Books,
c/o Pen & Sword Books Ltd,
47 Church Street, Barnsley,
S. Yorkshire, S70 2AS

www.greenhillbooks.com
contact@greenhillbooks.com

ISBN: 978–1–78438–495–1

Publishing History
First published in Germany in 2018 as
T-34: Russlands Standard Panzer im Zweiten Weltkrieg
by Motorbuch Verlag
Postfach 103743, 70032 Stuttgart
www.paul-pietsch-verlage.de

This is the first English-language edition
and includes a new foreword by Anthony Tucker-Jones

All rights reserved.

Original text © Motorbuch Verlag 2018
Anthony Tucker-Jones foreword © Greenhill Books, 2020
Translation by Geoffrey Brooks © Greenhill Books, 2020

The right of Wolfgang Fleischer to be identified as author of this work
has been asserted in accordance with Section 77 of the
Copyrights Designs and Patents Act 1988.

CIP data records for this title are available from the British Library

Designed and typeset by Donald Sommerville

Printed and bound in the UK by TJ Books, Padstow

Typeset in 11/14.5 pt Adobe Caslon Pro

Illustration credits: Archiv AKCM u.a. (78), Eiermann (1), Evers (14),
Fleischer (150), Hildebrandt (2), Jurga (11), Klotzsche (2), Muikku (7),
Töpfer† (1), Roth (1), Schweizer (2).

Previous page: A T-34/85 in the streets of Budapest during the suppression of the Hungarian Uprising in 1956. The T-34 had a long career in Soviet service.

Contents

Foreword	Stalin's Saviour *by Anthony Tucker-Jones*	vii
	Introduction	xiii
1	The Place of the T-34 in Tank History	1
2	Development	14
	Prototypes, Experiments and Series Production up to 1941	14
	The T34/76 – Mass Production until 1944	38
	Maintaining Mass Production and Making Improvements	107
	The T-34/85, 1943–1946	122
3	Specialised Variants	150
	Flamethrower Tanks	150
	Mine Clearance Tanks	159
	Recovery and Bridgelaying Tanks	161
	The T-34 for Active Propaganda	163
4	Self-Propelled Guns	164
	SU-122	165
	SU-85	174
	SU-100	184
5	T-34 Data	194

Foreword: Stalin's Saviour

It is a pleasure to write the foreword to Wolfgang Fleischer's study of the T-34 tank. The T-34 was one of the most remarkable tanks of the Second World War and remains a constant source of fascination. It is also one of the very few designs to be built from the beginning of the conflict until the very end. Only the German Panzer IV matched its longevity. The T-34 was a constant on the Eastern Front and proved to be Stalin's saviour. It subsequently enjoyed a long post-war career, seeing action in Korea and the Middle East. Along with the American-built Sherman and the German Tiger, it became a household name.

However, when the T-34 first went into action in the summer of 1941 it proved to be an unmitigated disaster. For a start there were not enough of them and the Red Army was left reliant on its more numerous, older and obsolete tank types. The first model T-34 was under-gunned due to the low velocity of its main armament. Then there were problems with the engine and transmission; there was insufficient ammunition and fuel, plus the crews were ill-trained. In the face of Hitler's Operation Barbarossa, T-34s were abandoned wholesale, often in very embarrassing situations, which bore testimony to the inexperience of their frightened crews. Others, lacking the skills to manoeuvre properly, were caught by German artillery and dive-bombers. Their tanks were blown apart or flipped like children's toys. Many T-34s became so much scrap on the road to Moscow. This outcome is graphically illustrated by the photographs in this book.

Hitler and his generals were initially not unduly concerned with the appearance of the T-34. Nonetheless, its radical sloping armour, wide tracks and 76.2-mm gun should have rung alarm bells. Then, in the winter of 1941, the Red Army, having learned valuable lessons, used the T-34 to maul the Germans at Mtsensk. Overnight the Germans realised they were in trouble. The T-34 rendered all the early panzers obsolete. Although the Panzer IV was upgunned, General Heinz Guderian demanded an immediate replacement. The subsequent Panther, which incorporated many of the T-34's features and was a fine tank, took too long to design and was never built in sufficient numbers.

In the meantime, the Red Army, having successfully evacuated its tank plants to the east, began to churn out thousands of improved T-34s, that had the early teething problems sorted out. General Georgi Zhukov dubbed this the 'Russian miracle' and the tanks began to arrive just in time for the defence of Moscow in late 1941. Frostbitten German troops soon found themselves being overrun by brand new T-34s supported by tough Siberian infantry. Major-General Friedrich von Mellenthin acknowledged, 'They played a great part in saving the Russian capital.' At the critical moment Barbarossa faltered.

The T-34's utilitarian and robust design made it easy to mass produce and with adequate training easy to use. In contrast many of the German panzers were over-engineered, costly and time-consuming to build. The T-34's distinctively sloped front glacis and sides offered greater protection and shot deflection than its German counterparts. Perhaps, though, among its greatest innovations were the wide tracks, which spread the weight of the tank and most importantly meant it could operate in the worst of conditions. Whereas the panzers often bogged down, the T-34 was able to pull itself out.

Ultimately it proved extremely rugged and reliable. For example, in 1943 the T-34 managed an operational readiness rate of 70–90 per cent. In contrast its rival the Panther managed just 35 per cent. The following year this had only risen to 40 per cent for the Panther and 60 per cent for

T-34s completed in 1940 (*nearest camera*) and 1941 (*centre left*). The inadequate training of Soviet tank crews on the new type was one of the causes of their heavy losses in 1941.

the Panzer IV. Interestingly the Red Army accounted for 75 per cent of all Panzer IV losses during the Second World War, with many of them destroyed by T-34s.

Equally important the T-34 was built with a turret ring that allowed for the installation of a larger turret and therefore a bigger gun. In early 1944 the T-34/85 appeared, armed with an 85-mm gun, which gave it parity with the heavier German panzers. Likewise, with the removal of the turret and installation of a box superstructure, the T-34 chassis was able to take 122-mm, 100-mm and 85-mm guns. The SU-85 and SU-100 proved excellent tank destroyers, although the interim SU-122 armed with a howitzer was not so good. Although the SU-100 proved nose heavy, its 100-mm gun was deadly against the Panther.

The T-34's finest moments were undoubtedly helping save Moscow, enveloping the Axis forces at Stalingrad, achieving victory at Kursk, then taking the Red Army to the gates of Warsaw and finally Berlin. One of the most famous formations to deploy the T-34 was General Pavel Rotmistrov's 5th Guards Tank Army. This fought at Kursk in 1943 and the following year during Operation Bagration, Stalin's answer to D-Day. At Kursk the only way to cope with the Germans' superior gun ranges was to close with them as quickly as possible. This resulted in the T-34 crews having to fight the panzers at point-blank range. Rotmistrov's appalling tank losses were such that Stalin almost sacked him. However, when Stalin learned that Prokhorovka had been turned into a giant panzer graveyard he rapidly changed his mind. The T-34 helped crush Hitler's last offensive power on the Eastern Front. This paved the way for Operation Bagration and victory.

Although the Red Army suffered continual heavy tank losses, it was able to replace them at a more than adequate rate. By the beginning of 1944 they were still able to field over 5,350 tanks and self-propelled guns. In the middle of the year, the Red Army deployed 2,715 tanks and 1,335 assault guns for Operation Bagration. Most of these were T-34/76s, T-34/85s and SU-85s/100s.

The T-34 was an immense success story, despite the best efforts of vested interests in the Soviet defence industries, who tried to thwart it from the start. Thanks to Marshal Grigori Kulik, in charge of the Artillery Directorate, the initial model T-34s were armed with an inadequate gun, even though a better version was available. Kulik was an old-school gunner, who had no time for tanks. He was of the view that artillery would simply pulverise them on the open battlefield. Kulik even tried to get Stalin to arm the T-34 with an ancient 107-mm field gun. The velocity of this would have been much too low to act as a tank gun, and would have required a much larger turret. Instead the standard 45-mm tank gun used on the BT-7 was upgraded to the 76.2-mm weapon, with a view to eventually increasing it to the 85-mm gun.

Mikhail Koshkin was the unsung hero responsible for overseeing the development and design of the T-34. He produced an almost impossible balance between armoured protection, firepower and mobility. No one had really achieved this before and no one was to match it during the Second World War. He did this by going out on a limb and designing a tank that went over and beyond Stalin's original specifications.

The events of 1941 forced large tank factories to be relocated to the Urals. Here, far to the east and behind the front, the fight for supremacy over the German Reich and its forces began.

Koshkin had been tasked to produce a successor to the series of BT fast tanks. These were too thinly armoured and their guns not powerful enough. The new medium tank requirement was designated the A-20. A heavier version known as the A-30 was quickly abandoned, but Koshkin pressed on with the A-20 and a side project known as the A-32 – this was to become the T-34. Stalin was so impressed that, although the design was incomplete, it was accepted by the Red Army in December 1939. Tragically Koshkin did not live to see the results of all his hard work in combat. He developed pneumonia while testing his creation on the road and died on 26 September 1940.

What follows is a translation of Wolfgang Fleischer's excellent book, originally published in German in 2018. This is a very welcome addition to the distinguished history of the T-34. He charts the evolution of not only the T-34/76 and T-34/85 in extensive detail, but also the myriad of different specialised support variants, such as the tank destroyers that sought to emulate the German *Sturmgeschütz* concept, flamethrowers and mine clearers. Importantly it provides a German perspective of what it was like to fight against this tank.

Four Polish brigades received T-34s from various manufacturers from 1944. The photo shows a tank of 1st Tank Brigade 'Heroes of Westplatte' in Danzig on 28 March 1945.

The book is lavishly illustrated throughout with a wealth of remarkable photographs (both archive images and of preserved examples) and technical drawings. It includes numerous photos new to me, especially of the relatively rare 1940 model. Ironically, although not many were built, most were lost during the opening stages of Barbarossa and the Germans took delight in photographing them wherever they found them. When they could the Germans employed captured T-34s and even made use of their turrets on armoured trains and bunkers! Of particular value is the technical evaluation, which provides a guide to the often less than obvious differences in the myriad of different models. These are supported by detailed plans of each type. Wolfgang had done a fine job and I commend this book to readers seeking to expand their knowledge of the T-34.

Anthony Tucker-Jones

Introduction

At first light on 5 March 1940 a convoy of motor vehicles set off from the grounds of the Comintern Locomotive Factory at Kharkov. Amongst them were two A-34 prototypes of a new medium tank developed for the Red Army, and two heavy tracked Voroshilovets towing tractors. Their destination lay 1,400 km away – Ivanovskaya Square in the Kremlin – where they were to be shown off to Josef Stalin, General Secretary of the Communist Party of the Soviet Union.

This journey was to herald the beginning of a stormy future for tank development in which the T-34 would have a lasting influence on the course of the Second World War. The predominant characteristic of the T-34 was the ideal combination of mobility, firepower and armour in a vehicle suitable for mass production. These advantages were brought to a level of perfection previously unknown in international tank design, such that at the outset of Operation Barbarossa, the attack on the Soviet Union, the T-34 was ready to cause major problems for German anti-tank defences.

Recent technical literature in the Russian language shows how, in contrast to work of the earlier period, the development of the T-34 branched out widely into numerous variants. It proved unrealistic to attempt to illustrate the multiplicity of prototypes, pre-series and series vehicles, self-propelled guns and specialised tanks in a book of this length, and the emphasis has been focussed accordingly on producing a well-balanced selection of models which determined the phenotype of tank and self-propelled artillery units in the struggle against Germany to the end of the Second World War, though an exception has been made for a few especially significant projects and experimental vehicles.

At this point it is customary to acknowledge the support given for the production of such a book. I express my thanks therefore to Dr Thomas Hug (Swiss Military Museum, Full), Igor Ballo (Bratislava), Lt.-Col. Marcel Fritz (Saumur Tank Museum), Thomas Evers (Elze), Rolf Hilmes (Ketsch/Rhineland), Konstantin Fromm (Bad Bergzabern) and Bernd Lausch (Dresdner Sprengschule GmbH).

It seems appropriate to remember a group active across the German Democratic Republic from the end of the 1960s who collected military materials in which the 'legendary T-34' naturally played a prominent role. The aim of this community of like-minded military enthusiasts was the evaluation of the meagre quantity of material available in order to give some shape to the officially released, one-sided, inaccurate and censored publications available behind the Iron Curtain, to preserve the knowledge that already existed, and to search for fresh information. In this regard I would like to mention Bernhardt Hildebrandt of Bad Doberan, Werner Klotzsche of Radebeul, the late Peter Meisel of Birkenwerder and Lothar Töpfer of Berlin. They all accompanied this author on the path which turned his hobby into a professional interest, and they deserve my special thanks here.

Wolfgang Fleischer
Freital, September 2018

1 The Place of the T-34 in Tank History

At 0315 hours on 22 June 1941, Germany invaded the Soviet Union by crossing its borders from the Carpathians to the Baltic Sea. After a very few hours Army Groups South, Centre and North were being made aware of a new and previously unknown Russian tank.

Two days later, on 24 June, new Russian tanks were discussed at the afternoon situation conference of *Oberkommando des Heeres* (OKH – Army High Command). The Chief of Staff, General Franz Halder, noted in his diary the following day:

> Details to hand about the new Russian tanks are: weight 52 tonnes, front armour . . . 8.8-cm flak apparently also penetrates at the sides . . . [a reference to the KV heavy tank – *Author*]. Another new type has been reported with a 7.5-cm gun and three machine guns [the T-34 – *Author*].

The T-34 came as a complete surprise to Germany's soldiers. At the beginning of 1942 a compilation of experiences of enemy tanks on the Eastern Front summed the problem up: 'The main strength of Russian tanks is not so much their armour generally as the slanted frontal plating. The outstanding manoeuvrability of the T-34 in the field requires special tactical measures, above all the use of terrain safe against tanks.'

V Army Corps headquarters stated in a report dated 15 December 1941: 'The confidence of the infantry in our own anti-tank measures has been

A burnt-out T-34 from the early production run at Factory No. 183 (Kharkov Locomotive Works). This photograph was taken in 1941. The T-34 presented German forces with major problems and once again laid bare the inadequacies of their anti-tank arsenal first exposed during the campaign in France in 1940.

shattered.' Panic often broke out when Soviet tanks were sighted advancing. On 23 March 1942, 88th Infantry Division Staff wrote in similar vein: 'The 3.7-cm [anti-tank] gun is inadequate for engaging those enemy tanks newly appearing on the battlefield.' From then on German soldiers gave the gun the ironic name 'the knocking-on-armour device' (*Panzeranklopfgerät*).

The T-34 worked as a catalyst in almost all areas of military technological development. In Germany the mass deployment of the T-34 quickly brought about a programme to modernise the panzers already in service and create new designs, amongst them the Panther and Tiger, and even the 190-tonne colossus Maus, at the limits of tactical and technical serviceability. Efficient self-propelled assault guns, anti-tank panzers, anti-tank guns and to some extent totally new anti-tank close-combat weapons figured in the response without successfully resolving the search for an adequate anti-tank defensive system. The head of Wehrmacht armaments, General Walter Buhle, summed it up at the beginning of 1945: 'All battles of recent times have provided a clear picture that the tank alone brought the real decision.' The T-34, of which tens of thousands were manufactured, swamped the German defences, and influenced further developments in other countries including France, Great Britain and the United States until well into the post-war period.

It stands to reason that from the beginning of its existence the T-34 would become an object of ideological and historical dispute. This persists into the present. An authority on the production of panzers and other military vehicles, Oberst Willi Esser, delivered judgement on the T-34 in a speech he made on 3 December 1942 to the Berlin working group of the *Wehrtechnische Gesellschaft* (Military-Technical Association) as a 'tank with strong firepower and very fast speed, a strikingly favourable specific ground pressure (about 0.70 kg/cm²) and a favourable form of armour

A German 3.7-cm L/45 anti-tank gun crushed by a Soviet tank. The gun crew stood little chance against a T-34. To be effective, even using the Pz.Gr. 40 round, a lucky hit was needed at less than 100 m, if possible against the tank's sides. Hence the ironic description of their weapon used by German anti-tank gunners: 'The knock-on-armour device'.

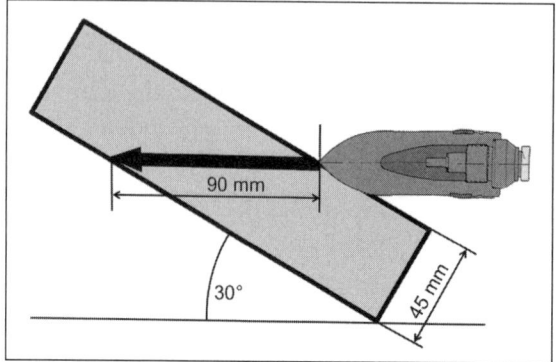

Diagram showing how sloping the armour at an angle of 30° doubles the thickness of armour to be penetrated by AP shot.

providing great protection against shellfire'. Esser was also emphatic in his summary:

> The Soviets began designing tanks more than ten years ago by the expedient of copying foreign tanks ... and in the stage of subsequent development almost slavishly incorporated many details and components of foreign origin, the result being that in design and construction, having regard to the circumstances of the Soviet Union, their tanks are remarkable and in many respects surpass those of our other opponents.

Karl Sedlatzek, author of the article 'What is the Value of Enemy Tanks?' published in the journal *Der Sieg* of 18 July 1943, described the T-34 as follows:

> Once the first weeks of the victorious advances in the struggle against the Soviet Union had petered out, there suddenly materialised before the German lines a tank which until then had remained hidden from the world. The Bolshevists had built it based on their several decades of experience copying foreign tanks, developing it specifically to suit the Russian climate and field conditions. This T-34 was actually ... a surprise against which, at the time of its debut, our available anti-tank defences were at a difficult stage ... equipped with a diesel motor, very broad tracks, forged steel plating and a 7.62-cm gun, it has an astonishing top speed of 54 km/h.

Oberstleutnant (Dipl.Ing.) Hans Albertz of OKH made a similar argument. In 1943 he described the T-34 as 'until now the most modern

The Germans were naturally interested in evaluating the Red Army's latest tank technology. This photo taken in July 1941 shows a captured T-34 being driven off by the Army Weapons Office/Weapons Testing Branch 6 to the testing grounds at Kummersdorf near Berlin. The T-34 was classified there as a 'medium battle tank 747 (r)'.

tank, and especially difficult to engage. It is being hurled into the battle in large numbers.'

Without doubt this fighting force, numerically strong in tanks, represented a threat. 'This danger has now been overcome,' wrote war correspondent Horst von Koblinski in his article of December 1941 entitled 'Steel fortresses turned into scrap metal', for they had been 'swept aside by the strategy of the *Führer* and the effectiveness of the anti-tank arm he has created'. However, the Soviet tank arm recovered from whatever the setbacks were, and the triumphant advance of the T-34 ended at the centre of Third Reich power, in Berlin.

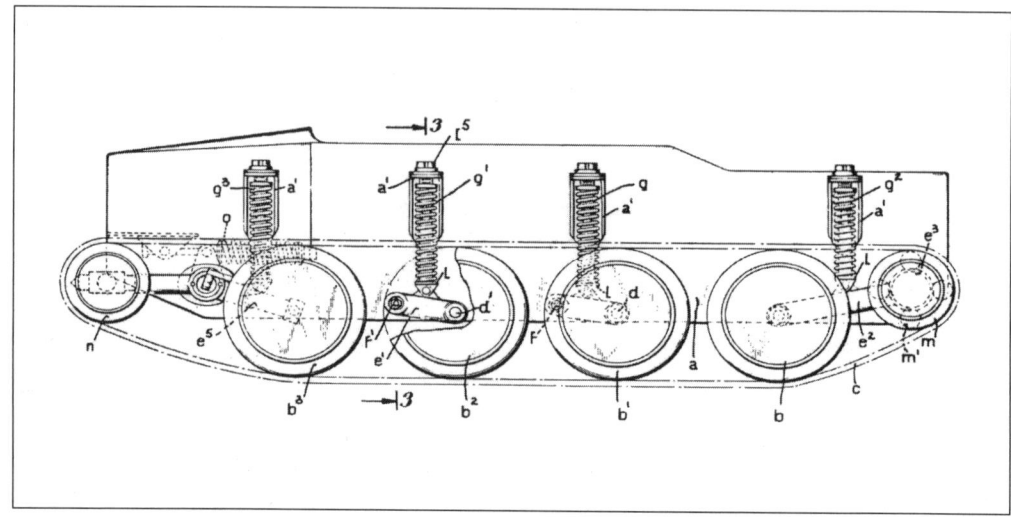

Drawing from the patent for the Christie wheel and suspension undercarriage, 30 April 1928.

How was that possible? The T-34, the result of a level of efficiency in the Soviet armaments industry never suspected in the West, requires explanation. In the search for answers, an anonymous author writing for the September 1943 issue of the *Wehrtechnisches Monatsheft* described the T-34 as 'a successful Christie-copy ... very fast and with noteworthy cross-country mobility'. After the war this opinion was shared by General Friedrich Maria von Senger und Etterlin, author of numerous technical works and articles on tank technology. The T-34 was for him 'the last limb of a development which began with the Christie Tank M in 1931'.

The T-34 – result of theft of intellectual property and plagiarism?

There is no word here about the quality of the tank, the well-synchronised combination of firepower, mobility and armour – achievements of the Soviet armaments industry in its own right. Even exiled technocrats in the era of the Cold War shied away from involving themselves in the depths of ideological argument regarding historical facts which were in very short supply at that time.

Other notable military experts with first-hand experience of the T-34 adopted an unequivocal stance from differing perspectives. Field Marshal Ewald von Kleist had risen to his high rank by 1943 after commanding a panzer corps, a panzer group and finally Army Group A. Prior to his extradition to Yugoslavia in 1945, in long conversations with Sir Basil Liddell Hart, in his time a famous author and theorist of modern military planning, Kleist delivered his opinion of the Red Army and its tanks: 'Its equipment was very good even in 1941, especially its tanks . . . Their T-34 was the best in the world.' The Soviet Marshal Ivan Konev, who held high positions of command in the Red Army in the Second World War, stated in his memoir *Sorok pyatyij* ('Forty-Five': Moscow 1966): 'The "Thirty-Four" survived the war from beginning to end, and no army had a better tank. No tank was a match for it . . . neither an American nor a German.'

It is interesting that Kleist and Konev – on opposing sides in the war – were to a large extent in agreement in their assessment of this tank type. Naturally Kleist took advantage of the stage provided for him by the British to contribute a reason to explain and justify the total defeat of Germany's armies in the field.

Konev's laudatory appraisal of the T-34 came twenty years later, in the 1960s at the height of the Cold War. His analysis was likely tainted by his desire to bring to the forefront recognition of supposedly superior Communist military policies and planning, and so the tank became incorporated into the argument used against the opposing political system. Therefore, even at the beginning of the 1990s, at the time of the collapse of the Soviet system, the development history of the T-34 was still shrouded in secrecy.

Details which could be made known were mistakes, shortcomings, errors and losses which awakened doubts regarding the infallibility of a leadership devoted to Marxist-Leninist ideology, but it remained simpler to sell the T-34 as the result of a perfectly legal home development. An admission that the idea, concept and design had been copied form the capitalist world would be going too far.

As late as 1966, the technical historian Colonel W. D. Mostovenko, an expert in Soviet tank technology, writing in the magazine *Technika i woorushenie*, was the first to admit that the wheel-or-track tank of the American J. Walter Christie had been the predecessor of the T-34. Thirty years later another expert, A. W. Karpenko, the author of 'Review of Russian Tank Technology 1905–1955', was still at pains to avoid making such a statement. Even famous military historians in East Germany had followed Karpenko earlier in their 1977 'Summary of the History of the Tank Weapon': 'The emergence of the Russian tank arm and the path

The wheel/track M.1931 tank designed by J. W. Christie seen running on its wheels.

it took forward prove again the superiority of the Communist military organisation which passed its greatest test of efficiency under the most difficult conditions on the battlefields of the Great Patriotic War.' The T-34, 'which for many years was the best medium tank in the world' came in for repeated mention as a shining example.

Polish tank expert Janusz Magnuski was less generous: for him the T-34 was only 'the tank of most renown in the world'. Specialist Western authors delivered a more impartial judgement. John Milsom wrote in his 1970 volume *Russian Tanks* that the T-34 was 'the best tank of its class at the beginning of it career . . . its achievements strongly influenced the later development of the tank everywhere in the world'. If this assessment is accepted, there is less danger of underestimating the military power of the Soviet Union.

The qualitative characteristics of the T-34 bear investigation from another perspective. Tanks are complex military machines. Their development requires planning and is influenced by many factors including one's own strategy and tactics and those of existing or possible future enemies, the state of one's science and technology, industrial efficiency, geography and climate in operational areas and the ability to operate and maintain such a demanding fighting arm.

Although the Soviet Union pressed forward its industrialisation hastily, this resulted in an efficient armaments industry – an important precondition for the enormous expansion of the Red Army between 1930 and 1937, and in parallel with this process operational principles were re-thought. General V. K. Triandafillov, Deputy Chief of Staff of the Red Army, formulated them anew in his work 'The Character of Operations of Modern Armies' published in 1929. As he saw it, future military conflicts would be fought as total wars by armed populations well organised in preparation for them and full of fighting spirit. Attack would be the preferred method of fighting and Triandafillov added; 'Nobody doubts any longer the enormous tactical significance of the tank.' Soviet military theorists came to the logical conclusion 'that a struggle against aggressive imperialist attacks on the Communist State will have to be fought with total commitment and ruthlessness until the aggressors are annihilated'. To strike decisively at the enemy in his own territory predicated tanks, many tanks.

The U.S. Army tested the Christie M.1931 tank at Fort Benning under the designation 'Medium Tank T3'. They bought seven examples for US$241,500 but decided against further purchases. The tank interested the Soviet military much more and they acquired a number through international trade.

The T-26 light tank was deployed in Spain during the Civil War under the designation carro de combate T-26 modelo 1933. This photo taken in March 1937 shows a T-26 captured by Franco's Nationalists. It was highly rated for its 45-mm gun, but the 15-mm armour proved inadequate.

By 1930, Soviet tank production had turned out the unimpressive total of ninety light T-18s (MS-1), a type with little future. In order to gain time for the development of more useful tanks, examples of designs were purchased from Britain and the United States, these having found little interest from the military authorities of these countries. They were copied, met the criteria and entered service with the Red Army as the T-27, T-37, T-26, BT-22, BT-5, T-28 and T-35. By 1935, annual production had reached 3,905 tanks and the continuously increasing numbers were used to set up four mechanised corps and six autonomous mechanised brigades, while more than a hundred other formations and units were equipped with tanks.

The Spanish Civil War (1936–9) provided the Soviet Union with the opportunity to test the viability of its tank technology in practice. By 1938 the Soviets had delivered to the Iberian Peninsula 281 T-26 light tanks and 50 wheel-or-track BT-5s. Here for the first time they came up against small-calibre special purpose guns such as the German 3.7-cm L/45 anti-tank gun, the shells from which could easily penetrate 29-mm armour plates at a range of 500 m and were too powerful for any tank the Russians could field in this theatre. 'In the contest between the tank weapon and anti-tank defences, the balance in Spain is inclined . . . decisively in favour of the latter,' Major E. W. Sheppard wrote in his 1938 book *Tanks in the Next War*. The shell had won against the tank – a fact which made clear to Red Army commanders at a stroke that, despite bulletproof and anti-splinter armour, all their tanks, including the T-28 and T-35 'breakthrough tanks', were obsolete. The hopes of achieving their objective of freeing Europe sooner or later 'from the chains of capitalism' had now receded.

In this connection – as a precursor to the clash between German and Soviet tanks in the Second World War – it is interesting to observe how the Germans assessed their experiences in the Spanish Civil War. The Army General Staff was of the opinion that no generalisation could be made. There had been no tank battles; in attacks there had been no in-depth staggering of panzer formations and the panzers supplied to the Nationalist forces had no armament heavier than machine guns.

On the international stage some experts were questioning the value of armoured fighting vehicles in future armed conflicts. The German General Staff would have no truck with such thinking. As a solution to the problem they envisaged massed attacks by panzers equipped with a gun able to overwhelm an anti-tank gun at a range of 900 m; greater mobility, and reliable support from other branches in the field (artillery, pioneers and so on). 'Only as a secondary measure can the superiority of the anti-tank shell be compensated for by making the armour thicker.' These were the outlines for the 15- and 18-tonne panzers developed in 1934–5 better known as the Panzer III (3.7-cm gun, Sd.Kfz. 141) and IV (7.5-cm gun, Sd.Kfz. 161).

The first series vehicles were delivered in 1936–7. At the same time the Soviets were working on new tank concepts with shell-proof armour. The first experimental vehicle, the T-46-5, had an armoured hull 60 mm thick in front. The Soviet military and armaments industry now saw the need not only to improve the basic fighting qualities of their tanks, but to find the

(*Left & below left*) Some of the many T-34s destroyed in action. Photos taken in winter 1942.

right blend of the greatest possible mobility, much greater firepower and shell-proof armour. In 1939–40, results were available from testing the T-34 medium tank and the KV heavy tank. While the KV still presented many technical difficulties to be overcome, the T-34 came out well. It was a universal tank for providing close infantry support and also suitable as a fast powerful tank with great range for far-reaching attack operations. This advance in tank design was one of the prerequisites to create for the Bolshevist nation the necessary military superiority to achieve its political goals. Was the T-34 the best tank in the world? Is such a thing possible? The answer to this question is left to the reader. This volume offers the facts.

In conclusion another comparison. Between 1940 and 1945, the Soviet Union built 112,500 tanks and self-propelled assault guns of which more than half – almost 59,000 – were T-34s.* This required a massive communal effort making extraordinary demands even of a socialist economy not organised on market principles. Why were so many T-34s built? Because so many were lost to enemy action, more than any other tank type during the Second World War.

---

* As a comparison, up to 1945, the armaments industry in the USA delivered 88,000 tanks, 49,234 of these being M4 Shermans. Between 1939 and 1945 Germany produced 24,700 panzers, 6,273 of these being Mk V Panthers (Sd.Kfz. 171).

2 Development

Prototypes, Experiments and Series Production up to 1941

Anti-tank defences were first developed to oppose enemy tank operations during the First World War. The best guns to deal with tanks had to be mobile, of 20–47-mm calibre and firing special armour-piercing ammunition. At a range of less than 1,000 m they quickly became the deadly enemy of tanks built with only bulletproof armour.

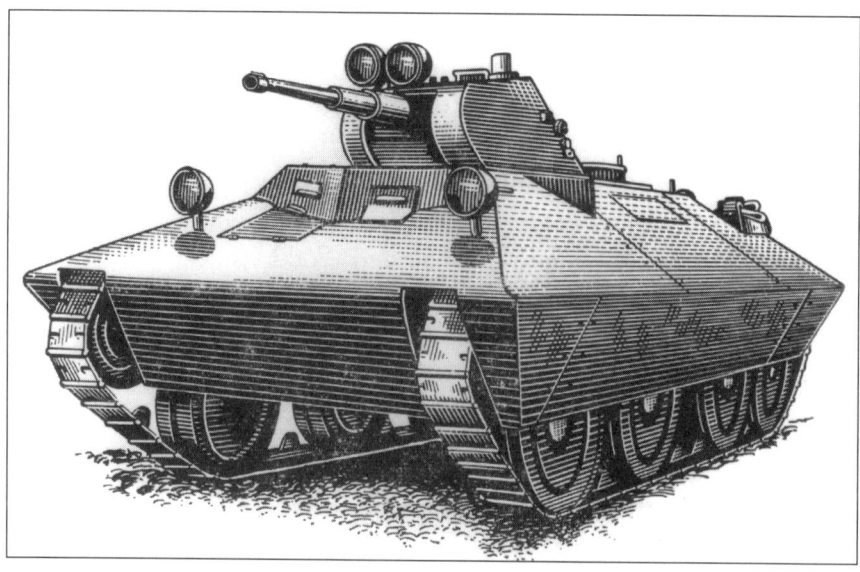

The hybrid wheels/tracks BT-SW2 'Turtle' was an experimental vehicle weighing 13.11 tonnes, 0.7 tonnes less than the BT-7 as a result of the sloped armour. The armour protected the crew against 12.7-mm MG fire (the BT-7 only against 7.62-mm fire).

The T-46-5 'Project 111' was only significant for Soviet tank designers gaining experience in the development of shell-proof armour.

The first use of anti-tank weaponry under war conditions occurred during the Spanish Civil War (1936–9). The T-26 and BT-5 tanks supplied to the Republican side were vulnerable to these guns, and it became clear at once that against well-organised batteries, there was no useful purpose to be served by putting tanks into the field with only thin bulletproof armour as protection.

The Soviet commanders recognised this problem and, based on their experiences, the Red Army's Administration Office for Mechanisation and Tanks (ABTU) devoted itself to designing tanks with shell-proof armour, greater firepower and – in comparison with BT tanks – not significantly reduced mobility. Greater prospects of success were thought possible from a new tank based on the Wheel-or-Track BT design which had greater performance potential than the barely 10-tonne T-26 equipped with a 90-hp engine. Experience in this regard was already to hand.

(a) Shell-proof Armour

In 1938, Repair Works No. 38 had produced a BT-5-IS with a new body form. The frontal armour provided a thickness equivalent to 26 mm slanted at 39°, the sides 13 mm slanted at 66° and the rear 13 mm slanted at 55°. However, this protection was only effective against projectiles up to 12.7-mm in calibre.

Subsequently Automobile and Tank Works No. 12 arranged a shell-deflecting slope to the armour plating on the BT-SW-2 wheel-or-track tank whose outstanding feature was its 6–25-mm armour inclined between 35° and 37° on all sides. The turret, with a 45-mm M.1934 L/46 gun and 7.62-mm DT MG, had armour 25 mm thick, curved at the front and inclined at 35° at the rear and on the sides. The peculiar outward appearance of the BT-SW-2 earned it the name *Cherepacha*, 'Turtle'.

The first Soviet tank with thicker armour came from the Leningrad Kirov Works (Factory No. 174) and was designated 'small tank with heavy armour T-46-5' (Object 111). The corresponding contract was awarded in the spring of 1937, and by October 1938 the factory had delivered two experimental vehicles which were tested in the February–April period of the following year. The top speed of 31 km/h and the range of 150 km on a full tank were deemed inadequate, and nor was the 8-cylinder 4-stroke MT-5 Otto engine with an output between 300 and 320 hp practical for a 32-tonne tank.

Progress had been made with the armour, however. Sixty millimetres thick at the front inclined at 70°, and 50–60 mm inclined at 72° for the turret, it guaranteed protection against armour-piercing rounds of 37-mm calibre fired at a muzzle velocity not exceeding 660 m/sec. Even 76.2-mm rounds did not penetrate from a range greater than 1,200 m.

(b) Significantly Higher Firepower

The armament of the T-46-5 was the 45-mm (1.8-inch) M 1934 L/46 in common use in Soviet tanks to that date, accompanied by three 7.62-mm DT MGs. A more powerful gun of greater calibre was required for a new tank to be mass-produced in the future. Experience with various types had already been gained.

From 1932 the 76.2-mm (3-inch) M. 1927/1932 L/16.5 (KT-28) gun had been installed in the multi-turreted T-28 medium and T-35 heavy tanks. A wheel-or-track BT-5 had been tried out with this gun in 1934. In 1936/37 a series of 154 BT-7As followed, classified as artillery support tanks, and in 1939 a 76.2-mm L-10 L/23.7 gun was installed experimentally followed by the L-11 L/30.5 and F-32 L/31 models. I. A. Machanov's Artillery Design Bureau I at the Leningrad Kirov Works (Factory No. 174) and Factory No. 92 (Gorky) led by V. G. Grabin were responsible for these projects.

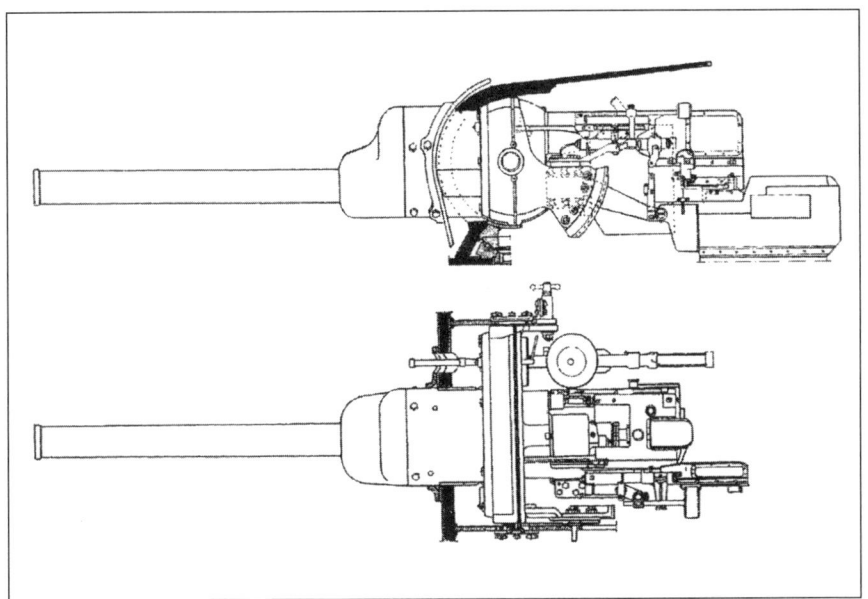

The 76.2-mm L-11 L/30.5 gun weighed 437 kg and had an exchangeable barrel lining with 32 grooves, barrel jacket, removable floor plate and semi-automatic drop-type breechblock. Recoil was between 380 and 470mm.

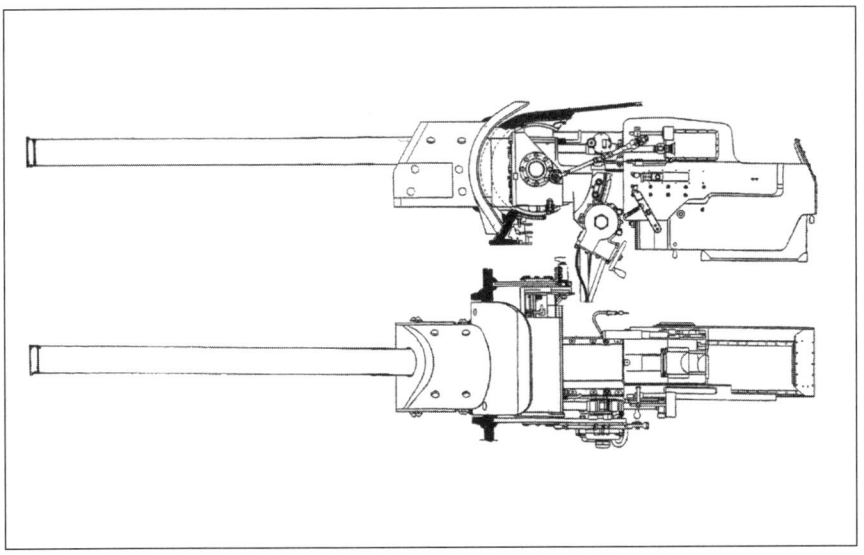

The 76.2-mm F-34 L/41.5 gun weighed 539 kg of which 470 kg was breech and barrel. The exchangeable barrel was similar to that described above. Recoil was between 330 and 425mm.

The transition to the longer-barrelled 76.2-mm gun brought about the required increase in firepower. Shells with a muzzle velocity of 555–612/613 m/sec penetrated 66 mm of armour plating from 500 m, more than twice what had been achieved by the 45-mm M.1934 L/46 (31 mm of armour).

(c) Not Much Less Mobility Than the Wheel-or-Track Tank BT

The power-to-weight ratio of the BT-7 was 29 hp/tonne. On tracks this provided the tank with a top speed of 52.3 km/h, remarkable for the time, and on bare wheels 72 km/h. The claimed range was 375 km (tracks) and 500 km (wheels).

Red Army exercises and manoeuvres in the 1930s confirmed the superiority of the tank in speed and range: a further increase in performance of the BT-7 was hoped for by replacing the 400-hp carburettor engine with a diesel (developed from the M-5 aircraft engine).

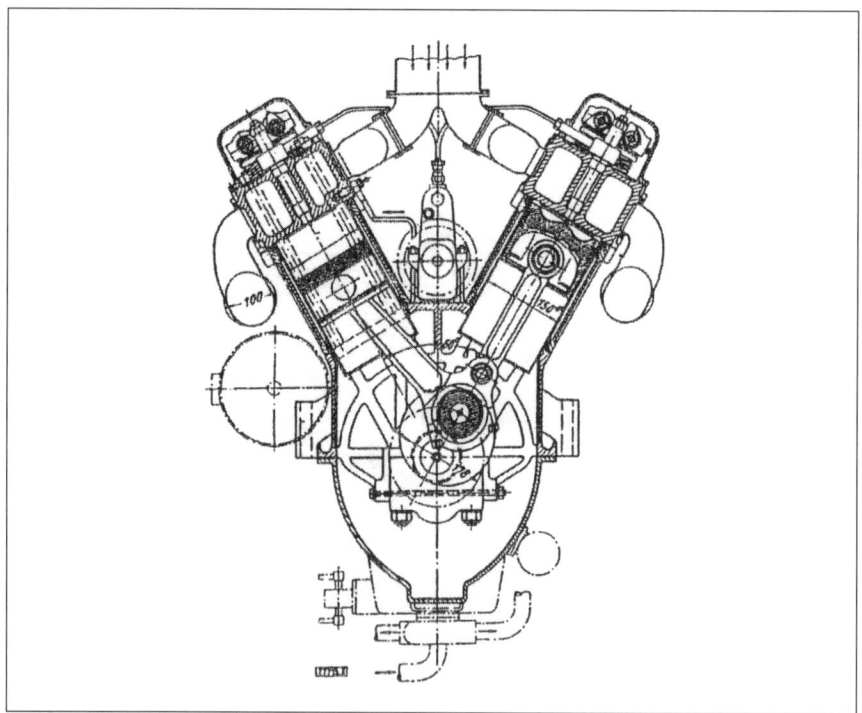

Cross-section of the 500-hp W-2 diesel engine. Weight 750 kg, length 1.558 m, breadth 1.15 m, height 1.072 m.

Between 1938 and 1940, Locomotive Factory No. 183 'Comintern' delivered 787 BT-7Ms (BT-8s) of which 715 were fitted with the W-2 diesel. This engine was still problematic and therefore 10 per cent of the 1940 output (72 tanks) received the 12-cylinder 4-stroke 400-hp Otto M17 engine (also designated BT-7m, M17). Photo from the summer of 1941.

The first trials in this direction had begun in 1931 with the AN-1 engine, worked on by designer A. D. Charomskim of the Central Institute for Aircraft Engine Building (ZIAM). In the years that followed, BT tanks were tested using high-output diesels specially made for tanks. The 12-cylinder 4-stroke 400-hp W-2 diesel was installed in the BT-7M (BT-8) manufactured by the Comintern locomotive factory at Kharkov which delivered 788 of the type during 1938–40. With a power-to-weight ratio of 34.1hp/tonne, on tracks the tank had a top speed of 65 km/h, and 86 km/h on wheels. A full tank of 778 litres of diesel fuel was sufficient for a BT-7M (BT-8) to cover between 630 and 1,250 km. Problems included damage to the armoured housing of the hull caused by vibration from the diesel, but this was outweighed by the advantages which included a good turning circle, low fuel consumption and reduced risk of fire.

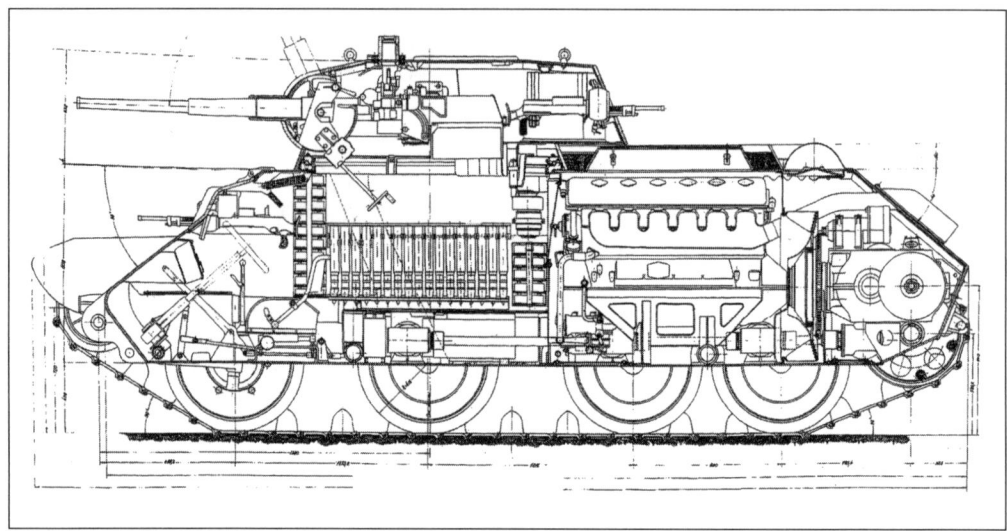

Project BT-20 was a predecessor to the new hybrid A-20.

To sum up: on the path to a tank with shell-proof armour, the T-46-5 was a poor development prospect. It had no reserve of power, the lack of space in the interior limited the size of gun which could be mounted, and nor was there room for the 1.45-m-long 12-cylinder diesel weighing 812 kg. It was not an option in the search for an infantry tank to replace the T-26, nor did it have the range to strike deep into the enemy rear.

Therefore the concept of a performance-increased BT tank was to be pursued, and on 13 October 1938 the ABTU office set out this specification for Project BT-20:

- Wheel-or-track tank with individual wheel suspension and three powered pairs of wheels;
- Armament:
    a 45-mm gun (ammunition 130–150 shells and cartridges) and three 7.62-mm DT MGs (2,500–3,000 rounds);
    a 76.2-mm gun (ammunition 50 shells and cartridges) and three 7.62-mm DT MGs (ammunition 2,500–3000 rounds): plus stabilised 'Orion' sight for range up to 1,000 m;
- 10–25 mm armour sloping on all sides;
- Maximum speed 70 km/h, minimum speed 7 km/h;
- Crew 3;

- Range 300–400 km;
- WD-2 diesel engine delivering 400 hp;
- Drive as for BT-5-IS with three powered axles for running on wheels.

The Comintern Locomotive Factory (Factory No. 183), which had provided the drawings and a model in March 1938, was awarded the development contract. The Development Commission had examined the trial results, however, and made a list of new requirements which led to the BT-20 being abandoned altogether during its project phase. The new tank design to replace it followed two lines of development from the outset: a wheel-or-track tank designated A-20 armed with a 45-mm gun, and the A-32 which ran on tracks only. This had either a 45-mm or 7.62-mm gun. The weight of both tanks was 15 tonnes.

In May 1938, the ABTU office issued the contract for the development of both designs to the Kharkov factory. Its chief designer was M. I. Koshkin supported by the KB-250 design bureau under A. A. Morozov. Three prototypes were to be manufactured as well as an armoured hull for testing purposes. The armour was to provide protection against 12.7-mm armour-piercing MG rounds able to penetrate 22.5 mm of armour inclined at 90° at 100 m.

For this purpose it was decided that rolled armour plates 20–25 mm thick fitted at the greatest possible angle of inclination would be sufficient. Engineer M. I. Tarshinov worked on the design of an all-round shell-deflecting armoured hull. Solutions of that kind had already been investigated with the BT-5-IS and BT-SW-2 ('Turtle') and these had determined the phenotype of the now-defunct B-20 project. Advantage was to be taken of the technological advances in electro-welding hardened steel made in the 1930s.

The 45-mm gun of the A-20, now in the M.1938 L/46 version, was retained as the main armament, accompanied by a stabilised Pribor 70 telescopic sight. Two 7.62-mm DT MGs completed the armament, one in the turret coaxial with the gun, the other in the front of the hull to the right of the driver. The complement was four men. The tried and tested 12-cylinder 4-stroke diesel, now the W-2 version, provided the drive with an output of 450 hp, later increased to 500 hp. The motor had four-speed transmission, a reduction gear being used to transfer power to the three

The A-20 tank, which could run on tracks or wheels, was presented in June 1939, a few months before the invasion of Poland.

Rear view of the A-20. The arrangement required to drive the three axles for running on wheels, and the two sprockets for running on tracks, proved too complicated for manufacture and maintenance.

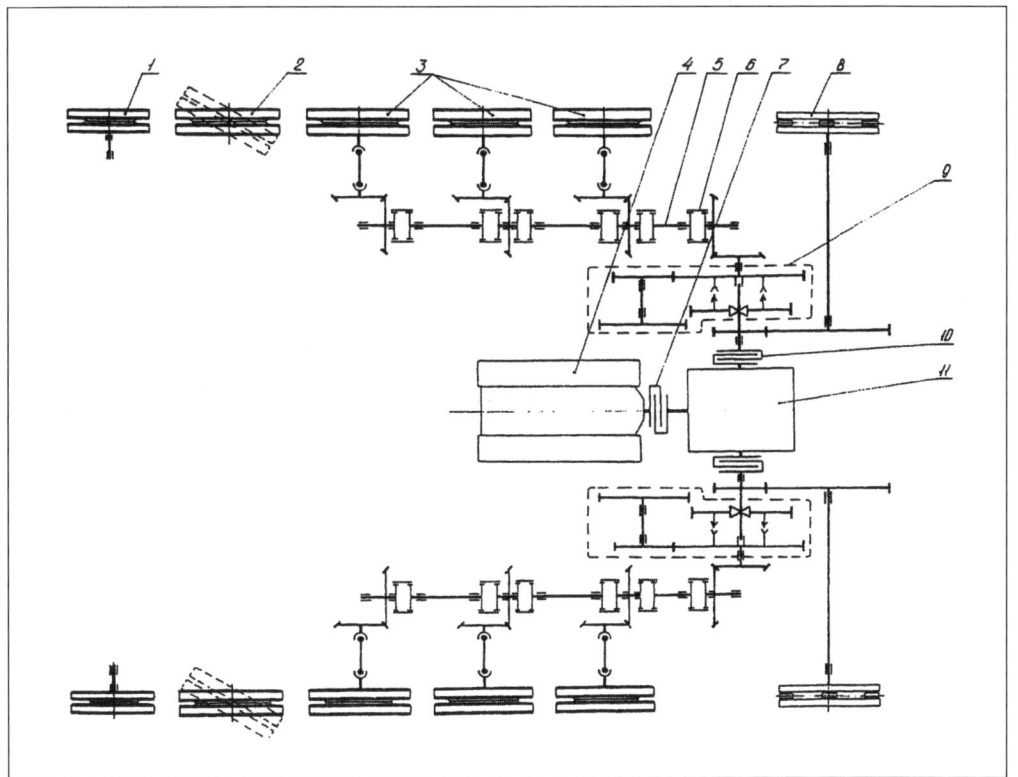

A-20 Drive Layout

Key: 1. Front idler; 2. Front road wheel pair for steering; 3. Road wheels; 4. Drive motor; 5. Drive shaft; 6. Gear wheel sleeve; 7. Main coupling; 8. Rear drive sprocket for powering the tracks; 9. Differential gear; 10. Lateral coupling; 11. Auxiliary gearbox.

---

axles when proceeding on wheels. The fourth pair of wheels was used for steering. The wheels had hard rubber tyres. During progress on tracks, power from the gearing passed by way of the lateral countershafts to the drive sprockets at the rear.

The speed of the A-20 on tracks was a maximum 65 km/h and 74.7 km/h on wheels. The A-32 drive was not so complex since the tank ran on tracks only, maximum speed being 65 km/h. Depending on the calibre of the main armament, its weight was between 18.3 and 19 tonnes.

In June 1939 the A-20 and A-32 prototypes came off the assembly line and were given runs on the manufacturer's own testing grounds where they were found to be reliable. Regarding the work, N. A. Kutcherenko of the Kharkov design bureau later stated:

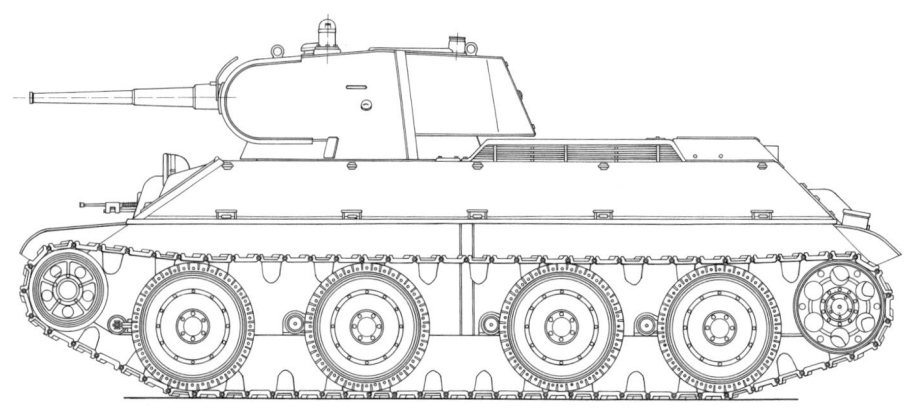

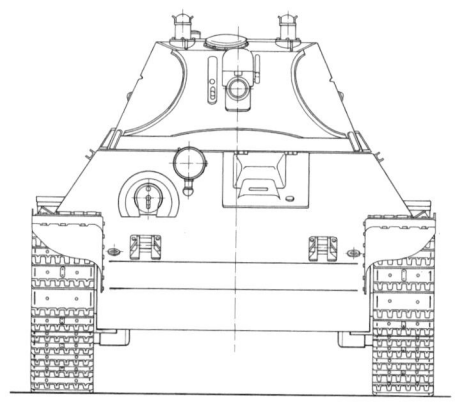

The A-20 tank. (*Drawings by Robert Jurga*)

Our design office had received the task of building . . . a tank . . . with both wheeled and track capability. This one was designated A-20. It turned out that its combat weight was overheavy by three tonnes which affected its speed and manoeuvrability adversely. We were all convinced that a tank of the specified weight should have a purely tracked chassis which would improve considerably the cross-country mobility without forfeiting speed. Removing the reduction gear for the wheels would make the tank lighter and free space for other fittings.

Despite the wishes of the would-be tank arm to have a wheels-and-tracks tank, there were two major factors which militated against it:

- The simplified design of the A-32 economised on valuable raw materials, saved a large number of ball bearings, and reduced the labour costs for the armoured hull;
- It also enabled an increase in armour protection above the existing levels resulting from the experience in this respect gained in the Spanish Civil War, which had finally ended in March 1939.

Nevertheless, on their first trial outings the A-20 and A-32 prototypes had shown adequate reserves of performance despite being overweight.

From 18 July 1939 further tests were made in the Kharkov region under the supervision of the chief of 1st Division ABTU, Major E. A. Kultshitsky. In parallel, the head of ABTU, General D. G. Pavlov, asked his senior engineer to investigate possibilities of increasing the armour on both vehicles without making changes to the basic design. It was calculated that a 5-mm increase all round on the A-20 would result in an increase of armour weight of 0.7–0.75 tonnes. For the A.32 an increase of 10 mm would add between 1.6 and 1.65 tonnes so that the operational weight would be a maximum of 19.6 tonnes (corresponding to a power-to-weight ratio of 25 hp/tonne.

As the international situation worsened, the Red Army Supreme Military Council gave fresh consideration to the question of how the tank arm should be equipped and armed. Marshal K. E. Voroshilov, People's Commissar for Defence, led the investigation in August 1939. Stalin also took part. It is interesting to note that at this time the military side was

The first version of the A-32 hybrid tank armed with the 76.2-mm L-10 L/23.7 gun. The five road wheels per side provided a more uniform distribution of weight giving a specific ground pressure between 0.61 and 0.55 kg/cm². In the summer of 1939 various weapons were tried out on board.

As an alternative, the second version of the A-32 was fitted with the 45-mm M.1938 L/46 gun. The armour approached that of the A-20.

not uniformly agreed on the future tank inventory. Some wanted fast dual-chassis tanks, others slower infantry tanks, and indeed the Leningrad Kirov Works (Factory No. 174) had already been given the order for an infantry tank to succeed the T-26. Those favouring fast tanks wanted the A-20; the A-32 was not popular. Koshkin, who had recognised the advantages of the tracked tank, came out strongly on the subject:

> I do not wish to go over the same old ground again, but I am of the opinion that for light tanks, travel on wheels is to a certain extent sensible. However, the complication of the dual-drive system for medium tanks is useless ballast, for then we have to provide the tank with more armour and better armament, and guarantee its manoeuvrability on the battlefield. The T-32 project [or A-32] on which we have worked is distinguished by the following characteristics: Regarding speed it is not inferior to the fast light tanks. It has the armour of the former medium tanks [T-28] and the armament of the medium and heavy tanks [T-28, T-35 and KV]. Accordingly it is a kind of universal tank and inferior to none. It is superior to the light tanks in armament and armour, to the medium tanks in armament and to the heavy tanks in speed and manoeuvrability. It seems that it can be used operationally as a 'breakthrough' and infantry-support tank as well as for independent operational tasks.

A decision to continue the development of both tanks was taken after heated discussions in which even Stalin spoke out. A few days later the Defence Committee referred the matter to the Council of People's Commissars. By now it was the end of August and the second A-32 prototype, armed with a 45-mm gun, had been delivered. According to the plan its handling trials were to be completed by 5 September, but finished earlier. On 5 September all three tanks were loaded up and sent to the tank proving grounds at Kubinka.

The trials team was pressed for time but had their final report on Voroshilov's desk by 20 September. It favoured the A-20 once more, its armour, engine, operational and fighting qualities being emphasised. After 4,200 km of travel, defects had materialised in the clutch, drive and braking systems. Poor visibility from inside the tank, especially for the driver, also came in for criticism. The manufacturer was therefore ordered to remedy

Characteristic features of the A-34 were the armour, increased to 45 mm, and the 76.2-mm L-11 L/30.5 gun.

the shortcomings and by 1 January 1940 another fifteen A-20s had been completed and preparations put in hand for mass production.

The two A-32s had been driven 3,000 km in their trials. The weaknesses coming to light were similar to those of the A-20, but on the positive side it was stressed that the greater reserves of power would enable an increase in armour thickness. Meanwhile 45 mm of armour had been demanded, sufficient to provide certain protection against 37-mm anti-tank shells. General D. G. Pavlov issued the corresponding contracts to Kharkov for calculations and stress testing at the end of August, requiring the factory to deliver five A-32s with the increased armour by 1 January 1940.

On 22 September 1939 the Red Army and industry presented the results of two years' development work at Kubinka to the Soviet leadership. They were shown the T-26 infantry support tank completed in 1938 and now modernised with sloping armour: the BT-7M and A-20 tanks, both capable of travel on either wheels or tracks, the tracked A-32 tank, and prototypes of the T-100, SMK and KV 'breakthrough' tanks.

A conference was held at the People's Commissariat for Defence and Heavy Industry in Moscow next day attended by the competent ministers, factory directors and senior designers. During the course of the meeting, problems regarding the preparations for series production of the A-20 to be undertaken in 1940 were clarified. Delays could be foreseen on account of its complicated design. Therefore, having regard to the deadlines, the A-32 represented the best option though with 45-mm armour and a 76.2-mm gun. Over the next two months, ABTU laid down other requirements. The A-32 – now being redesignated A-34 – was to be equally as manoeuvrable as the A-32, while the main armament was to be the 76.2-mm F-32 L/31 gun and three machine guns.

On 25 November 1939 the tank was given the final designation T-34 and the factory at Kharkov received a contract for the delivery of two tanks by 15 January 1940, another ten by mid-September and 200 more by the year's end. For the following year 1941 ABTU wanted a thousand T-34s from Kharkov and another thousand from the Stalingrad Tractor Works

Second version of the A-32 had a 'driver's window' with exit hatch in the hull roof. To enable the driver to leave, the turret had to be rotated aside first.

(STS), which had now been drawn into the series production. As a bridging solution a thousand BT-7M (BT-8) tanks were ordered. It was planned to cease production of this type in 1942, but in the event it actually concluded in 1940.

The overwhelming pressure exerted by the Soviet leadership on industry is shown by the decision of the Council of People's Commissars and the Central Committee of the Communist Party on 5 June 1940 to accelerate the manufacture of the T-34 into mass production in the current year 1940 as shown in the table. A comparable increase in production was expected in the manufacture of the W-2 diesel at Factory No. 75 in Kharkov where 2,200 of these were to be completed by the year's end at the rate of 210–350 per month. These ambitious targets had no hope of being met. The first series-produced T-34, classified as a medium tank, rolled out of the work hall at Kharkov in September 1940. Stalingrad produced none. Major difficulties occurred in the organisation of series production there regarding which the Central Committee of the Communist Party demanded a report.

The prototype of the T-34.

Planned T-34 Production, 1940								
	June	July	Aug.	Sept.	Oct.	Nov.	Dec.	June–Dec.
Locomotive Factory Kharkov (No. 183)	10	20	30	80	115	120	125	400
Stalingrad Tractor Factory (STS)	–	–	–	–	20	30	50	100
Totals	10	20	30	80	135	150	175	500

The operational weight of the T-34 had now risen to 26.8 tonnes and the power-to-weight ratio had dropped to 19.1 hp/tonne. Top speed was 54 km/h. The ground pressure was slightly increased (from 0.606 kg/m^2 to 0.62 kg/m^2, compared with the German Panzer III at 0.99 kg/m^2, and the British Matilda II at 1.12 kg/m^2) as a result of the half-metre wide tracks, one reason for the excellent cross-country mobility of the tank.

There were also changes to the armament. Originally the 76.2-mm L-10 L/23.7 gun fitted to the A-32 was to have been replaced by a gun of the same calibre developed at Gorky and designated the F-32, but its development and manufacture had not kept pace with the demands and this forced the decision to arm the T-34 with the improved 76.2-mm L-11 L/30.5. The F-32, intended for the KV heavy tank, remained unavailable and so recourse was had to the L-11 for the KV as well.

Without doubt the Soviet tank builders achieved a major advance in quality with the T-34, but the haste with which its development had been forced through resulted inevitably in problems which included the following:

- The inadequacy of having four gears for an engine of that output and the poor gear-change facility with corresponding high wear-and-tear on the clutch.
- The neglect of ergonomic considerations caused a shortage of space in the interior of the tank. The hull was comparatively narrow on account of the width of the tracks: the plating was vertical below and sloping inward above: the coil spring system continued from the BT series, though giving excellent suspension, helped limit the interior, and having the entire drive unit at the rear also required a great deal of space.

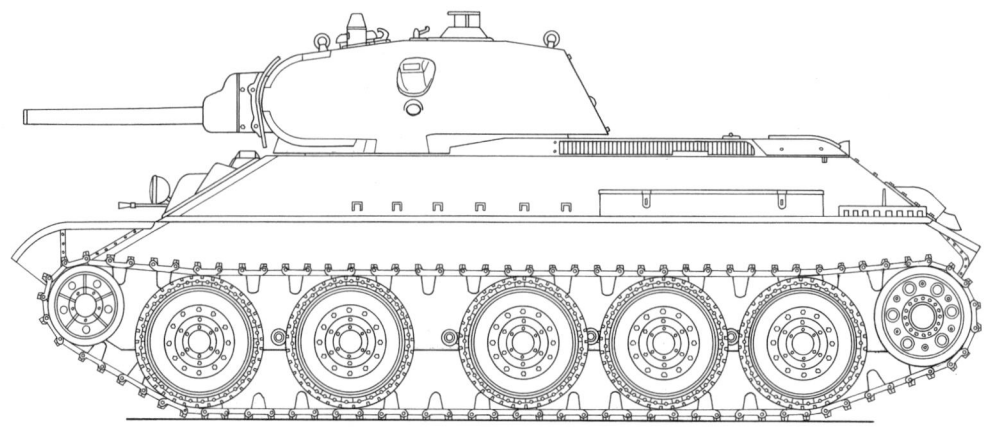

T-34, 1940 model. (*Drawings by Robert Jurga*)

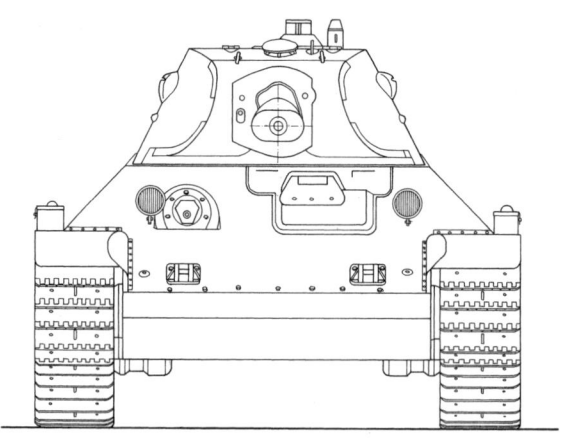

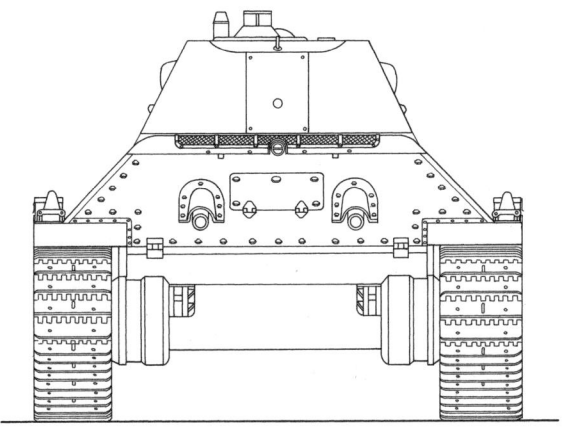
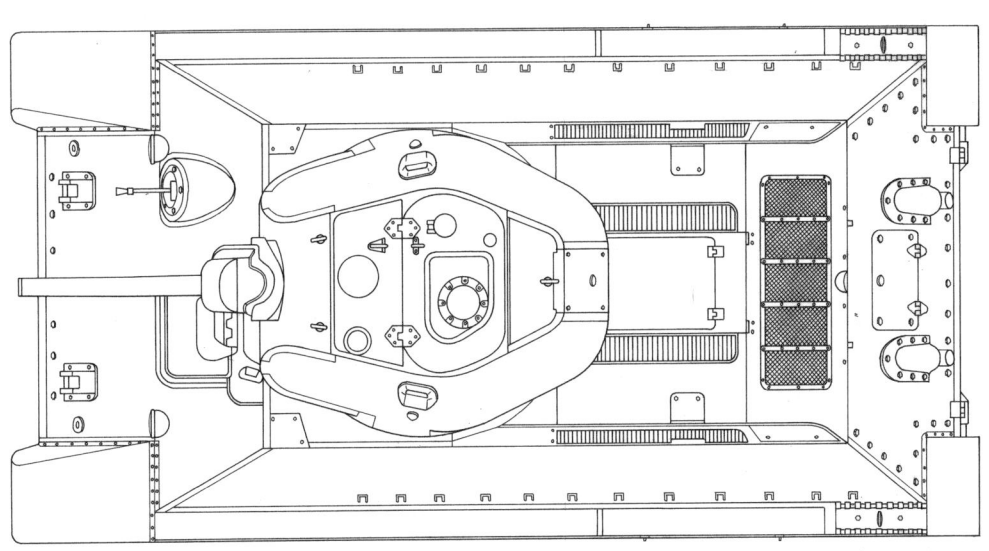

T-34, 1940 model from the series run at the Kharkov Locomotive Factory.

The first T-34 series run left the production line from the Kharkov factory in September 1940. Its existence was a closely guarded secret, kept even from serving Soviet troops.

- The turret with a ring diameter of 1.42 m could only accommodate two men, the loader and tank commander, the latter also having the role of gunner, which meant that he was overburdened and, since there was no commander's cupola, his view from the tank was unsatisfactory and limited his tactical appreciation.
- Three exit hatches (turret, driver and emergency exit through the floor) were insufficient.
- The reduced speed of 54 km/h as compared to the 70 km/h of the A-32 was thought to be insufficient. Other hopes for improvement were expressed with regard to enlarging the magazine and equipping T-34s in general with radio. The 20-watt 71 TK3 set (wavelength 75–53.3 m, 4,000–5,625 kHz, range up to 5,000 m) was not issued to any T-34, however.

At about this time a new tank project, conceived as the successor to the T-34, was taken in hand by designers at the Kharkov Comintern Locomotive Factory under the designation A-34. Its characteristics were as follows:

- A new 12-cylinder 4-stroke W-5 600-hp diesel to be installed at right-angles to the direction of travel.
- Combination of the transmission with a reduction gear (eight forward and two reverse).
- Undercarriage with torsion rod springs, six running and four support wheel pairs without hard rubber tyres. Together with the good suspension, this design provided more space in the interior and an increase in ground clearance to 550 mm.
- Driving seat repositioned on the right-hand side.
- From the design bureau of the Lenin Works at Mariupol, Ukraine, a hexagonal circular-track turret of 1.7-m diameter, commander's cupola and two exit hatches; the turret could accommodate three of the five crew members of an A-34.
- Transfer of the radio set from the turret to the hull. The advantageous distribution of space made possible the enlargement of the magazine from 77 to 193 rounds for the gun and from 2,394 to 4,536 rounds for the two MGs.
- Installation of the more efficient 76.2-mm F-34 L/41.5 gun.

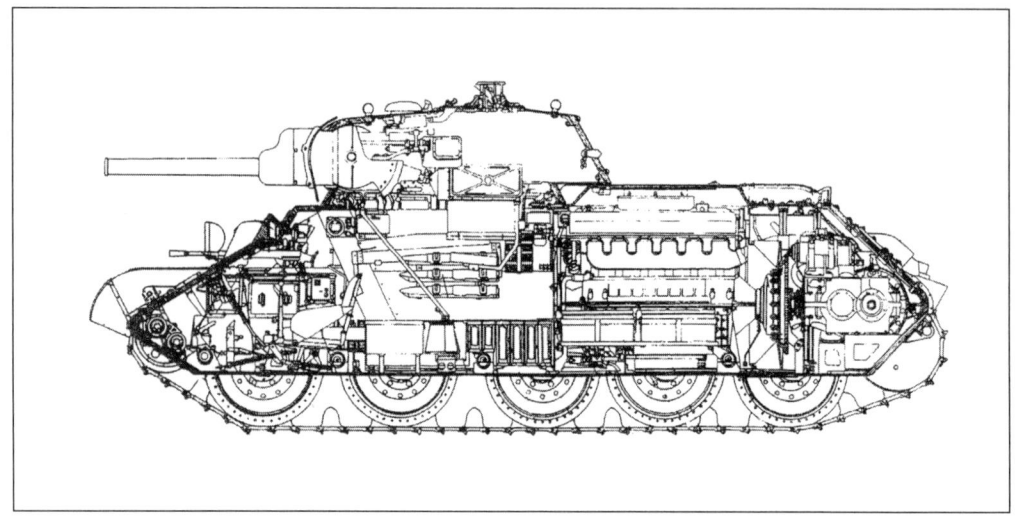

Cross-section of a T-34.

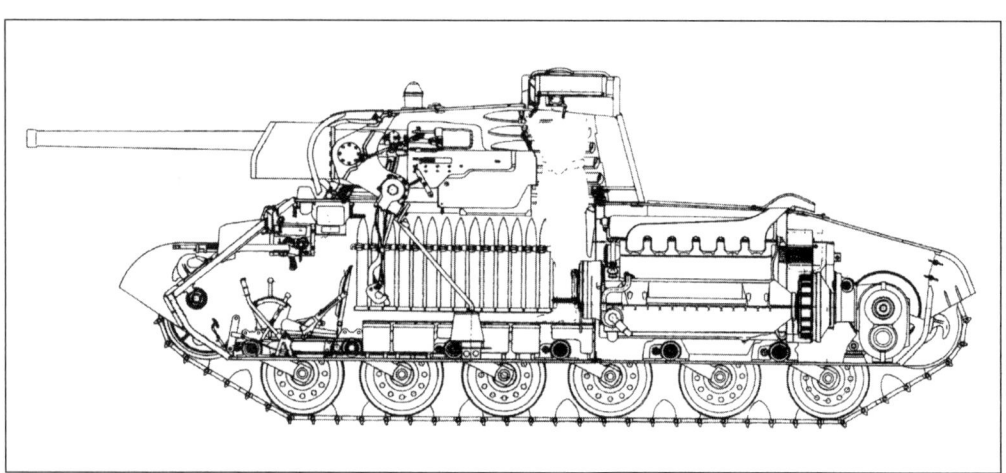

Cross-section of the A-34 with W-2 diesel engine.

- Shell-deflecting armour 45 mm thick; other options (60 mm, 75 mm, 90 mm and 120 mm) were examined.

Initially the Kharkov engineers drew up plans for a 25.5-tonne tank with a top speed of 60.5 km/h. Increasing the armour resulted in a weight increase from 29 to 32 tonnes and a cut in the top speed by 11 km/h.

Production Planning A-34, 1941							
	July	August	Sept.	Oct.	Nov.	Dec.	Total
Numbers (target)	10	100	175	180	165	170	800

Only drawings and models are known of the final A-34, In April 1941 three hulls were made at Kharkov and three rotating turrets at Mariupol. On 5 May 1941 the Central Committee of the Communist Party decided on series production of the improved tank designated T-34M with 60-mm frontal armour (battle weight 27.5 tonnes) The Committee ordered 500 units for the current year and later increased the number to 800 (*see table*).

A flamethrower tank was planned as a variant of the T-34M. The Lenin Works at Mariupol delivered fifty turrets in the summer, beginning in parallel with the production of turrets with 52-mm frontal armour for the series production of the T-34, itself now well under way. The series

1940 model T-34, from that year's delivery of 115 tanks. Seen here clearly is the form of the turret front with cast mantlet for the 76.2-mm L-11 L/30.5 gun.

Wooden model of the T-34 (also designated variously as T-34M in the literature), April 1941.

production of the W-5 diesel encountered major problems, however, and finally fell by the wayside.

The principal reason for none of these tanks being assembled was the abrupt change in the military situation. On 22 June 1941 the German attack on the Soviet Union began, forcing the Red Army to abandon large areas of territory in the west of the country and transfer its industry to the Urals and Siberia. Under these conditions great efforts were required to maintain series production of the T-34, while increasing the numbers and improving the existing models.

The T34/76 – Mass Production until 1944

Between 1920 and 22 June 1941, the Soviet armaments industry delivered 30,120 tanks of all kinds, 1,244 of which were T-34s, 115 manufactured in 1940 and 1,129 in the first six months of the following year. By 1 June 1941 the military administration had accepted 891 of them. Most went to tank units in the west of Russia where from 22 June the heaviest fighting against German forces occurred along a front 1,400 km long. In the course of it the Red Army tank force suffered very heavy losses.

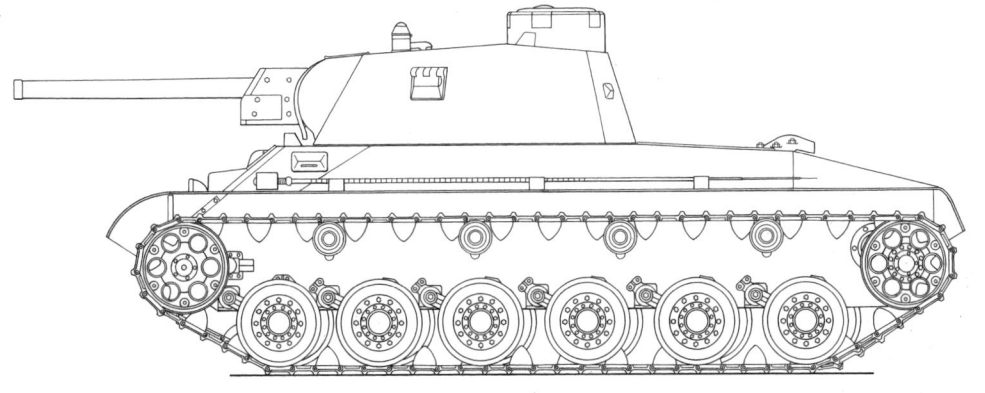

The A-34 (T-34M) medium tank. (*Drawings by Robert Jurga*)

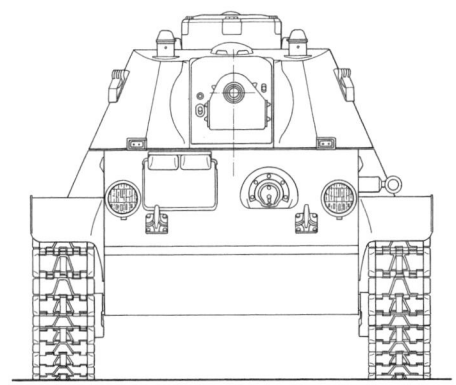

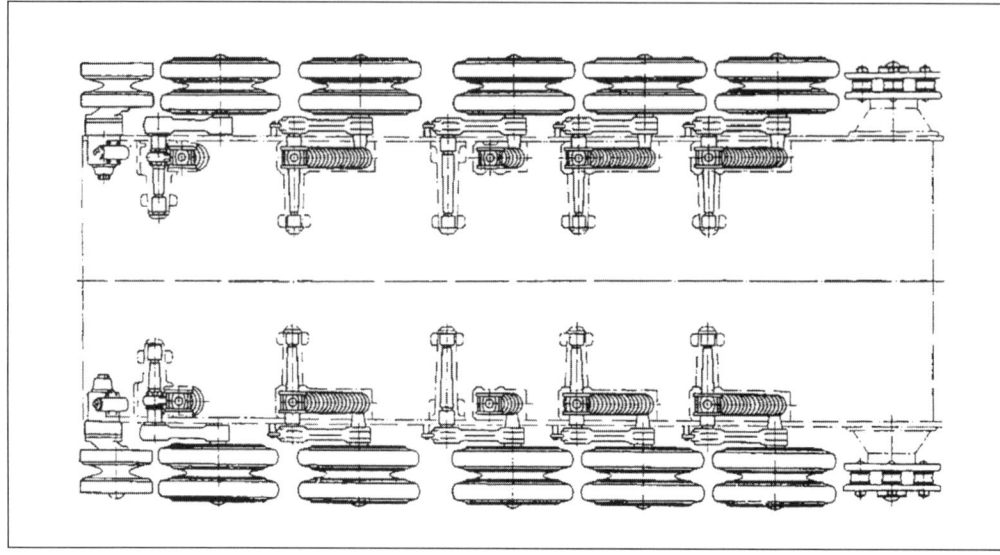

Diagram of the T-34 wheel plan showing the arrangement of the coil springs.

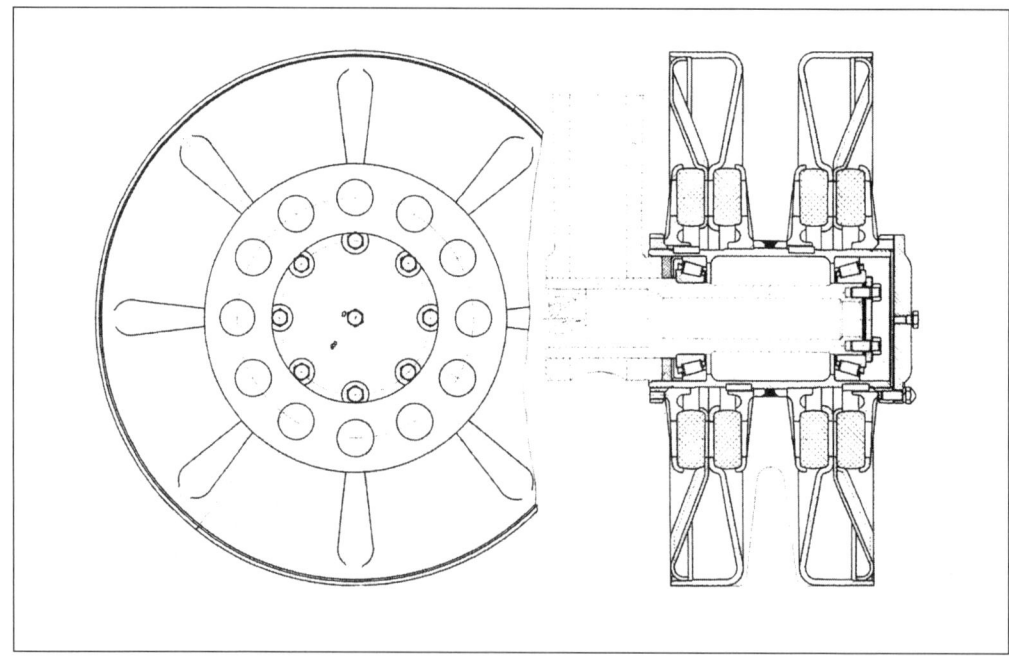

Technical drawing of the steel road wheel with rubber inlay as envisaged for the T-34. The shortage of rubber resulted in this solution for the T-34 in the winter of 1941.

1940 model T-34, with cast turret and 76.2-mm L-11 L/30.5 gun as delivered from Kharkov with effect from February 1941. Note the panoramic periscope on the turret hatch.

The German Army's Foreign Armies East intelligence department reported to the Quartermaster-General at Army High Command (OKH) on 10 October 1944 in retrospect that the Soviet tank losses in the second half of 1941 had been 22,000. This was in contrast to the 7,400 tanks which the German intelligence services had reported as newly built, a number not far from the reality. According to Soviet sources, the true production figure had actually been 5,984, of which more than half were T-34s. Production in other countries during 1941 was Germany 3,257; Great Britain 4,841; USA 4,052.

During this period the Soviet armaments industry was struggling: production was stagnating and in some areas even in decline. In order to improve the situation for the Red Army, especially to compensate for the enormous wastage in tanks on the battlefield, special efforts were needed. A massive increase in the output of tanks was required and so new factories for the T-34 and the W-2 diesel were opened in the east of the country to guarantee continuing production. This was carried out during the severe winter of 1941, with some production work even taking place in the open.

The first tanks were delivered after seven weeks, man with defects and low in quality. Upon examining wrecked or captured T-34s, German engineers took note of the sub-standard welding of armour plating and shortages of equipment. Kharkov had come under immediate German threat from the second half of September 1941 and the city was occupied in mid-October. News from there had never been promising and T-34 production had already been abandoned. On the other hand, the Stalingrad Tractor Works had overcome its difficulties and for the first time began delivering appreciable numbers in 1941. Therefore of the four factories which had received contracts for the series production of the T-34 before the German invasion, only three were still operating.

The efforts of designers and technicians at the surviving factories were concentrated primarily on raising output and maximising quality, improving the existing models, and simplifying the design and manufacturing techniques. In the years following, changes were made to the turret, armoured hull, armament, undercarriage, drive installation and

Seven of the 115 T-34s manufactured in 1940 were fitted with a radio in the turret and another five with a radio in the hull. Photo from July 1941.

Production of the Model 1941 T-34		
Factory and location	Quota	Delivery
Locomotive Factory (Comintern) Works No. 183 Kharkov	1,800	1,873
Factory No. 183 Nizhny Tagil	?	25
Factory No. 112 Gorky	?	161
Factory No. 174 Omsk	?	0
Uralmash Sverdlovsk	?	0
Stalingrad Tractor Factory	1,000	958
Total	2,800*	3,017
* Information incomplete		

other equipment. The inclusion of changes to remedy problems into the series production of the T-34 armed with the 76.2-mm gun (subsequently designated the T-34/76) was always ongoing until production was terminated in 1944. By its very diversity, it would be very difficult to arrange all this data into some kind of chronological record. Over this period, differences on the assembly lines were agreed between the six manufacturers of the T-34 and their sub-contractors.

Turret

The turret consisted of several welded parts: six plates of rolled armour steel, the cylindrical mount with cast mantlet for the 76.2-mm L-11 L/30.5 gun and a large single-piece hatch in the roof. The 76.2-mm F-34 L/41.5 gun had a mantlet made up of several parts. This was simplified in 1942 to a pattern from Factory No. 264.

The design of the turret was complicated and held back the expansion into mass production. The problem was caused by having the turret front curved above and below, and the homogeneous 45 mm rolled steel plate interlocking into the side walls with a piece left out for the gun mount. T-34s of this design were delivered by Factory No. 183 from the summer of 1940 until towards the end of 1941, and then finally from the new factory at Nizhny Tagil. Others came from the Stalingrad Tractor Works in 1941. They were replaced by tanks with a simplified turret of partly interlocking plates, though the interlocking helped improve the resistance to shell hits. The armour was 52 mm thick all round.

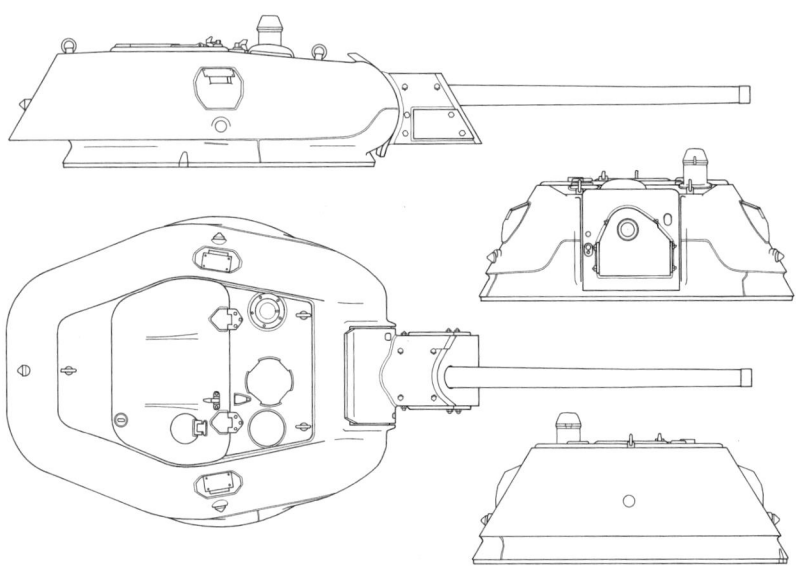
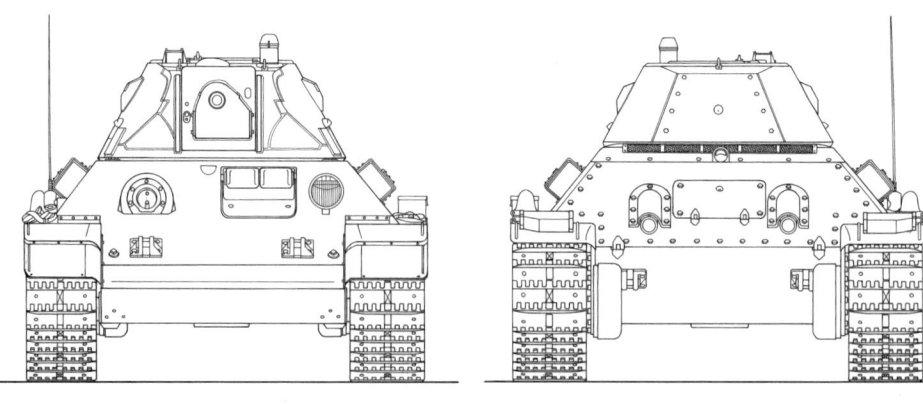

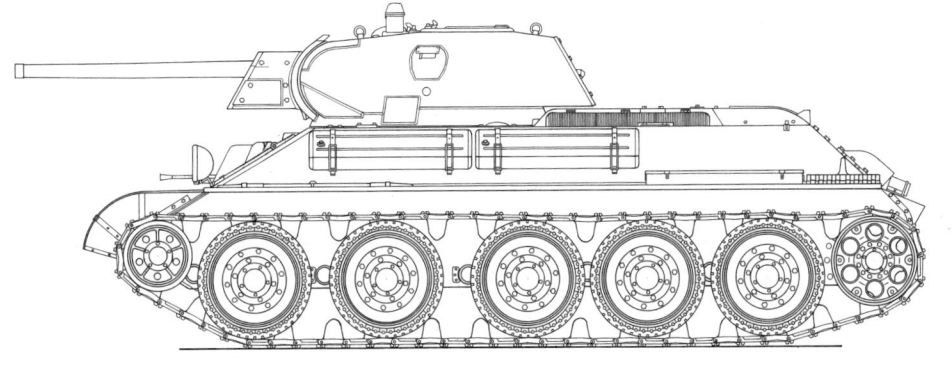

1941 model T-34. (*Drawings by Robert Jurga*)

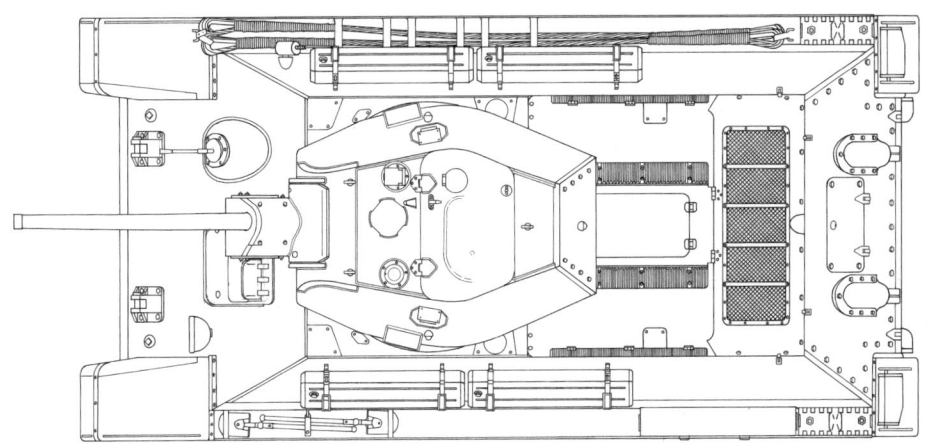

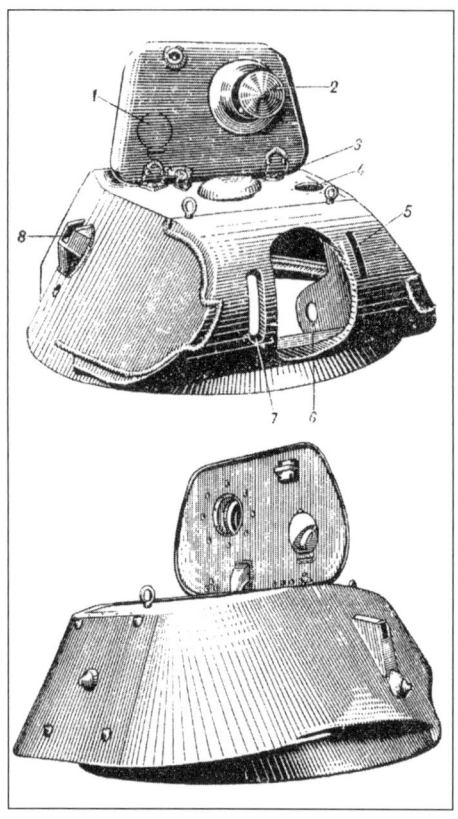

Front and rear view of the 1940 model T-34 turret: *Key:* 1. Signal flag hatch; 2. Panoramic observation device; 3. Ventilator; 4. Aperture for PT6 periscope (× 2.5 and 26°); 5. Slit for TOD-6 rectilinear telescopic sight (× 2.5 and 15°); 6. Fitting for gun trunnions; 7. Aperture for coaxial 7.62-mm DT MG; 8. Loader's viewing port.

The Lenin Works at Mariupol began the manufacture of the new turrets in the summer of 1941. In subsequent months the work had to be farmed out to Stalingrad and Magnitogorsk where the resumption of production was tied to rationalisation. At the end of 1941 Stalingrad produced a turret design seen as a provisional solution: only the upper part had the curved plating, two triangular pieces fitting into the turret housing below it. Everything possible had been done to strengthen the armour. The turret housing received an additional 15 mm, but not everywhere. In February 1942 the Stalingrad factory delivered 109 tanks with thicker armour, then in March eight more with turret walls 75 mm thick. These were integrated into the series then current, but were not included in the mass production run.

During 1942, Stalingrad turned out T-34s with turrets produced by other manufacturers, but despite all

Photo showing detail of the rounded interlocking frontal plate, a typical characteristic of the earlier version of the T-34 turret. The 76.2-mm F-34 L/41.5 gun is fitted.

rationalisation measures assembly proved expensive. Efforts were made to overcome this difficulty. In the second half of 1942, Factory No. 183 began deliveries of T-34s with a hexagonal turret. The parts were manufactured partly pressed and welded automatically under flux. Characteristic features were the forward-facing cast steel turret housing and welded roof armour with two circular hatches. The lateral viewing ports with armoured covers were replaced by simpler versions: some individual builders retained the pistol ports, enabling them to be closed from inside.

The new turret weighed 4.32 tonnes. The Chelyabinsk Kirov Works went about turret construction another way with pressed or completely cast turrets, the former being very noticeable for their angular form, the latter for their rounded shape. As was the case with all other T-34/76 tanks, from the spring of 1943 these were delivered with a cylindrical cupola for the commander, this feature increasing the height of the tank to 2.52 m from the previous 2.4 m. The comparatively small turret-ring diameter of 1.420 m, which was a result of the shell-deflecting form of the hull, subsequently proved inconvenient as more powerful guns were fitted.*

The introduction of the hexagonal turret and the elimination of the large one-piece hatch meant that the gun could not be removed from the tank without lifting off the turret. At first, since the A-34 design, there had been a 45-mm-thick removable plate at the turret rear screwed on from inside. This was still being used in 1942 on T-34s from Factory No. 183, but in order to simplify the manufacture of the turret side walls with their curve at the rear, in 1942 Stalingrad decided to screw this maintenance access plate across the width of the turret, giving the rear an angular appearance. Ultimately the maintenance access hatch was also abandoned, first by Stalingrad, the other manufacturers following suit later. On these tanks, the gun could only be removed through the large hatch in the turret roof, a complicated procedure. The transition to the hexagonal and cast steel turret from the Uralmash Works made even that impossible. As on the later T-34/85, the turret had to be lifted off and put aside in order to remove the gun. Another change was the repositioning of the ventilator from above the breech of the gun to the rear of the turret.

---

* The turret ring of the German Panzer III had an outer diameter of 1.675 m (barrel recoil of the 5-cm L/60 gun was 32 cm).

Model 1940, 1941 and 1942 T-34s.

The differences in armament and turret shapes are illustrated.
Below, bottom & right: T-34, 1940 model with 76.2-mm L-11 L/30.5 gun until the beginning of 1941, from Factory No. 183, Kharkov.

Model 1940 T-34 with 76.2-mm L-11 L/30.5 gun in the cast turret from the 1941 run at Factory No. 183, Kharkov.

All four: Model 1941 T-34s with 76.2-mm L-11 L/41.5 gun in a turret from an early run (curved and interlocked frontal plate, manufacturer Factory No. 183 Kharkov.

Both above: Early Model 1941 T-34s with 76.2-mm L-11 L/41.5 gun made by Factory No. 183 Kharkov).

Both above: Model 1941 T-34s with 76.2-mm F-34 L/41.5 gun in the cast turret. Manufacturer in 1941 was Factory No. 183 in Kharkov, which from the beginning of 1942 was at Nizhny Tagil.

Above: Model 1941 T-34 with 76.2-mm F-34 L/41.5 gun in the cast turret.

Above & right: Model 1941 T-34s with 76.2-mm F-34 L/41.5 gun from Factory No. 11 Gorky. The turret does not have the removable armoured plate at the rear.

Above, left & below left: Model 1942 T-34s with 76.2-mm F-34 L/41.5 gun from the Stalingrad Tractor Factory. Characteristic features are the turret front and rear, the latter having the broad removable armour plate.

Both above: Model 1942 T-34s, with 76.2-mm F-34 L/41.5 gun in the hexagonal turret with separate hatches for commander/gunner and loader and PT-4 and PT-K periscopes, from Nizhny Tagil. This version weighed 30.5 tonnes, that from Factory 174 Omsk, 30.8 tonnes.

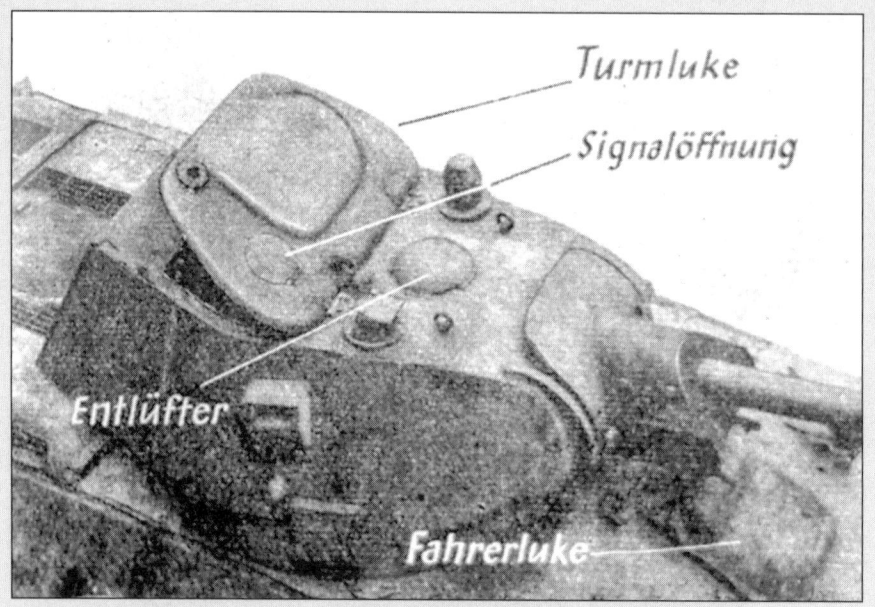

Model 1942 T-34, with 76.2-mm F-34 L/41.5 gun in the hexagonal turret. *Turmluke:* Turret hatch; *Signalöffnung:* Signal flag hatch; *Entlüfter:* Ventilator; *Fahrerluke:* Driver's hatch.

Above, below left & below: 1942 model T-34s, with 76.2-mm F-34 L/41.5 gun in pressed steel hexagonal turret. Manufactured by the Uralmash factory, Sverdlovsk, which supplied only 719 tanks of this type in 1941/1942.

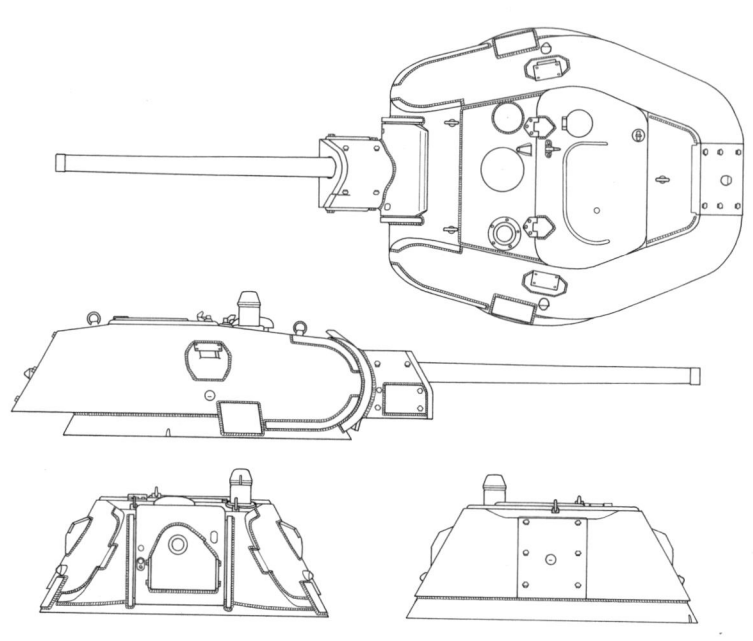
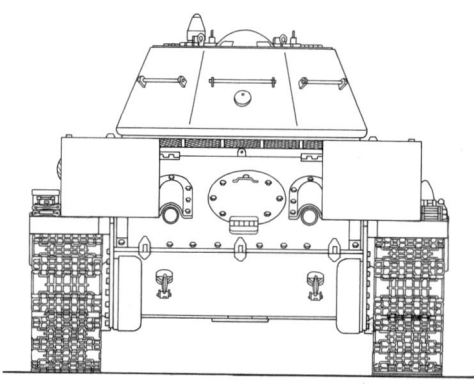
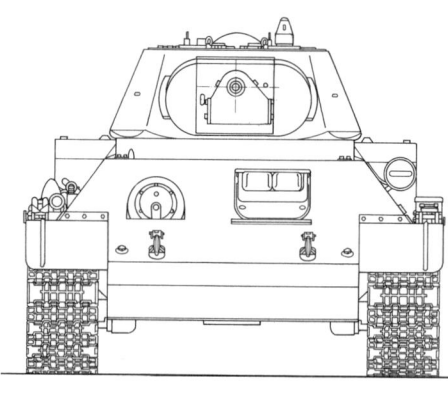

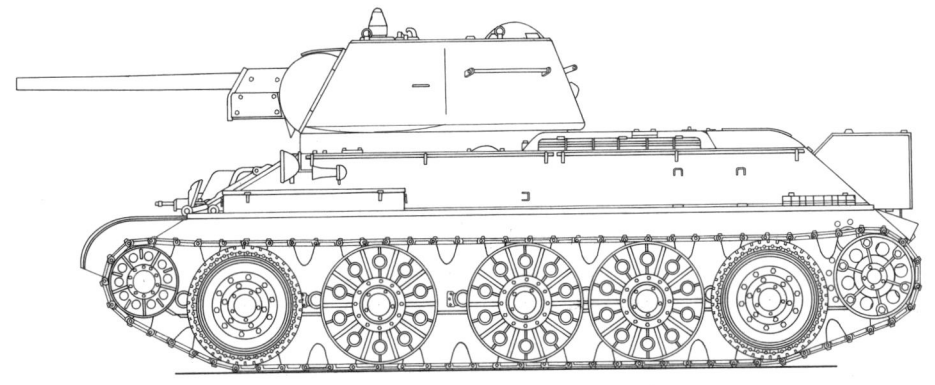

1942 model T-34. (*Drawings by Robert Jurga*)

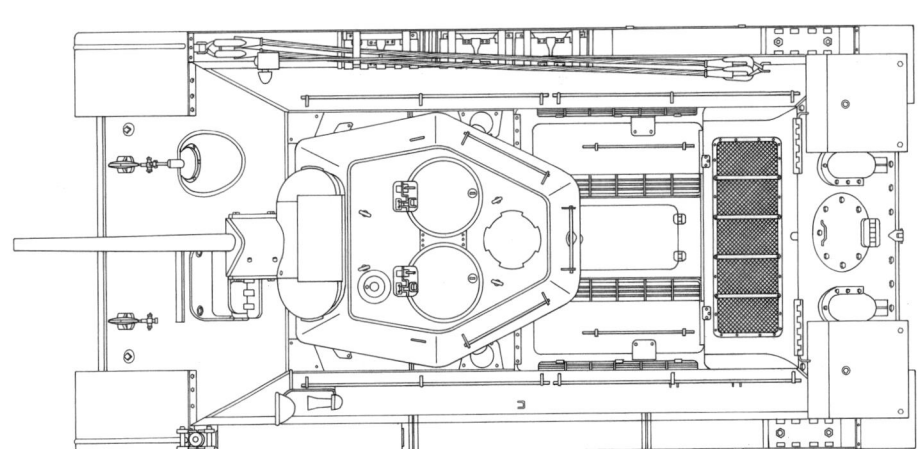

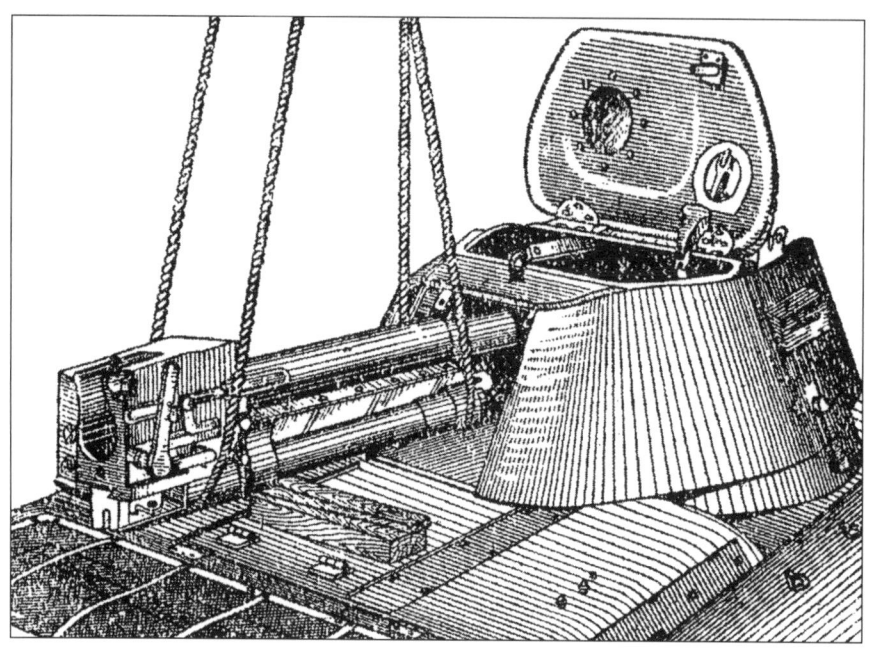
Removal of gun from turret with removable rear plate.

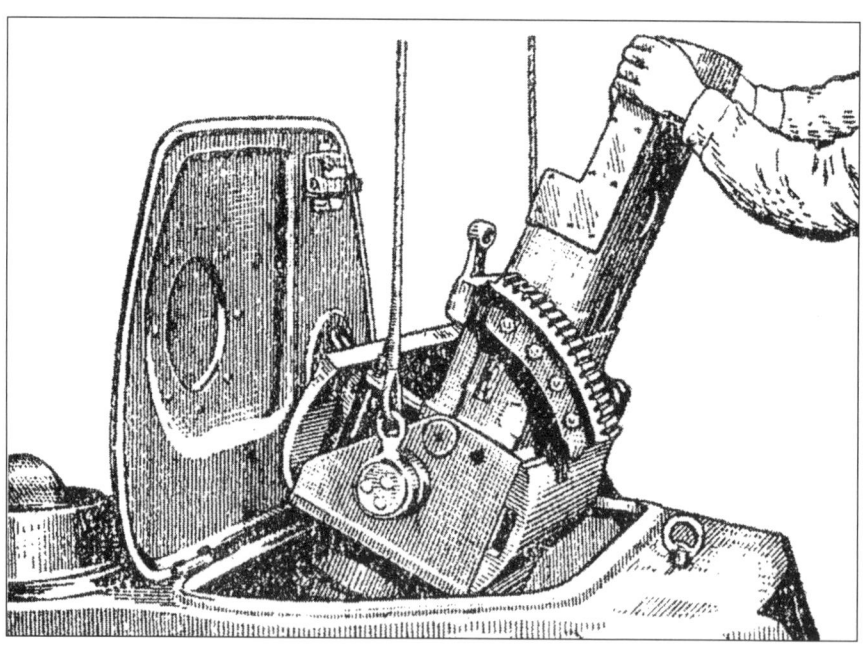
Removal of gun from turret without removable rear plate.

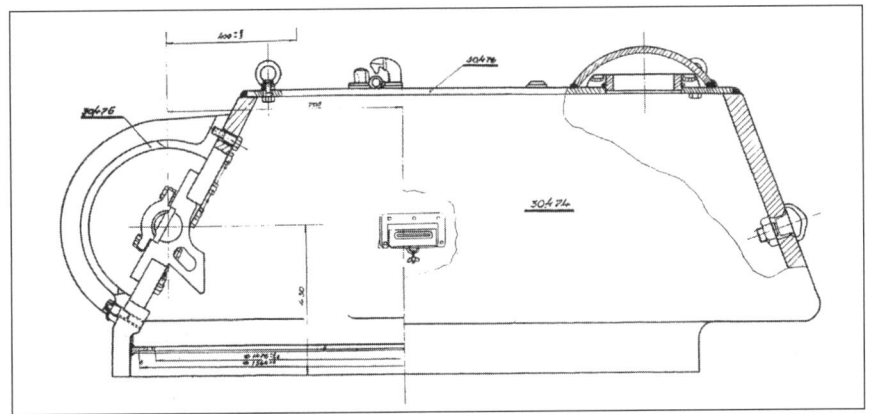

Drawing of the hexagonal turret from Factory 183 Nizhny Tagil, at the state of development reached by the beginning of July 1942.

1942 model T-34 with 76.2-mm F-34 L/41.5 gun in cast turret completed at Uralmash, Sverdlovsk. In 1942 the Chelyabinsk Kirov Works delivered turrets of this kind with a commander's cupola.

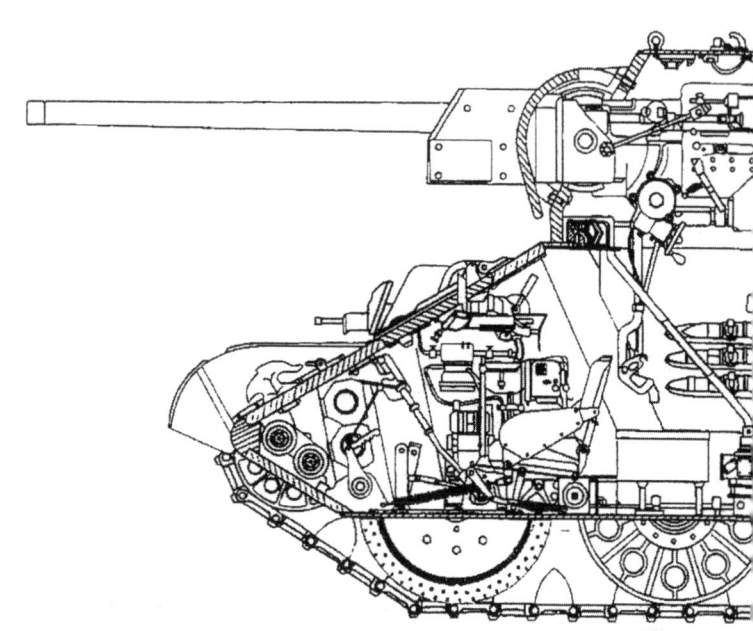

1942 model T-34 with cast turret from the Chelyabinsk Kirov Works, wrecked by enemy action, spring 1943.

Cross-section of a 1942 model T-34.

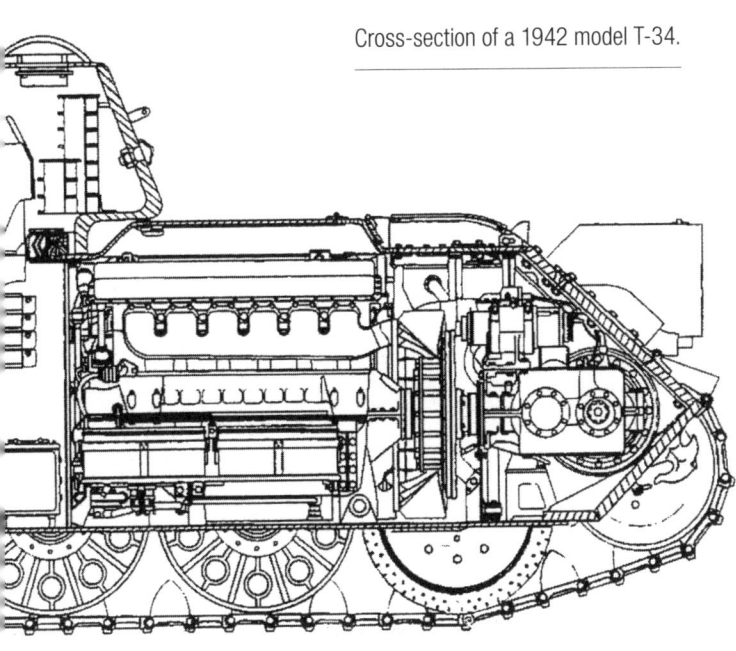

The T-34 turret delivered in 1943 by Factory No. 112 Gorky had minor differences.

1943 model T-34 with 76.2-mm F-34 L/41.5 gun with commander's cupola fitted, manufactured from April 1943 at Uralmash, Sverdlovsk.

1943 model T-34 with 76.2-mm F-34 L/41.5 gun. From the spring of 1943 the turrets had a commander's cupola fitted. Tanks with this type of turret were delivered by the Nizhny Tagil factory.

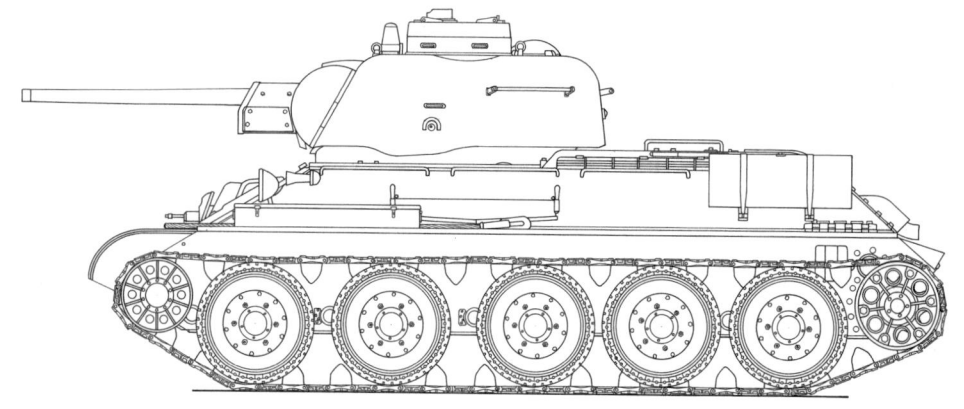

1943 model T-34. (*Drawings by Robert Jurga*)

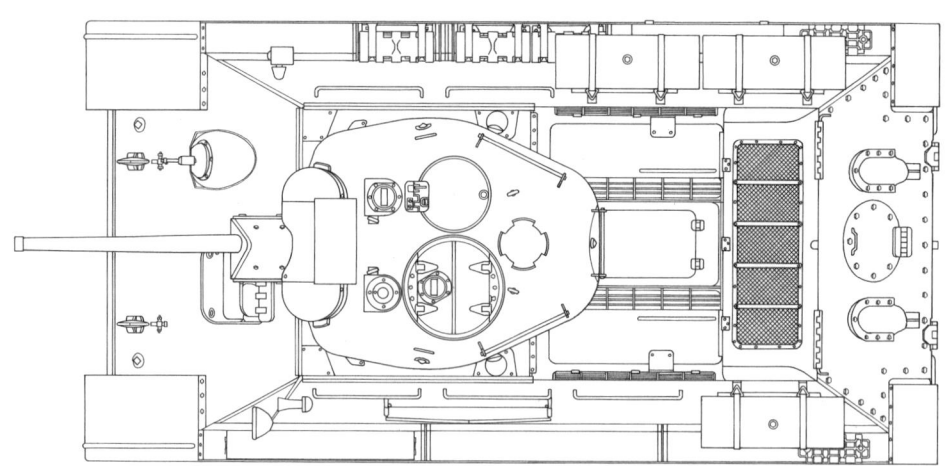

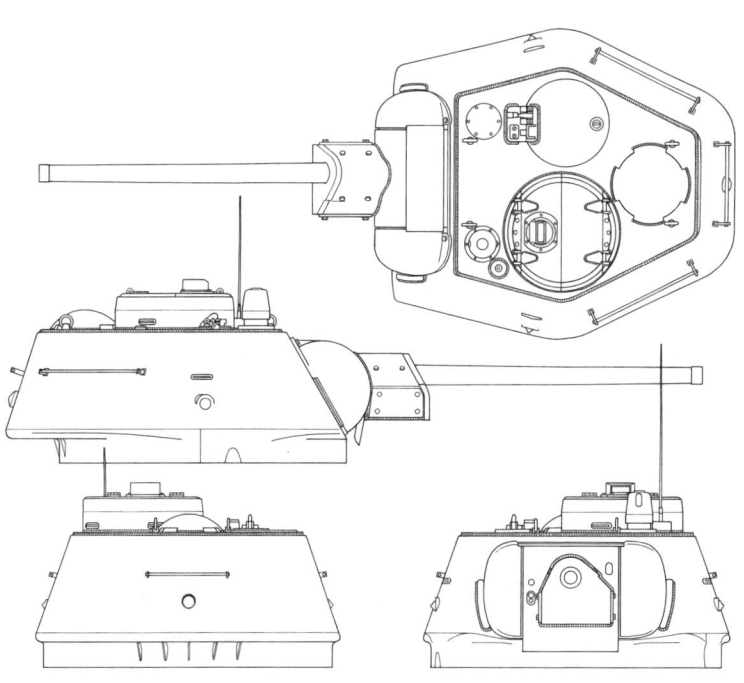
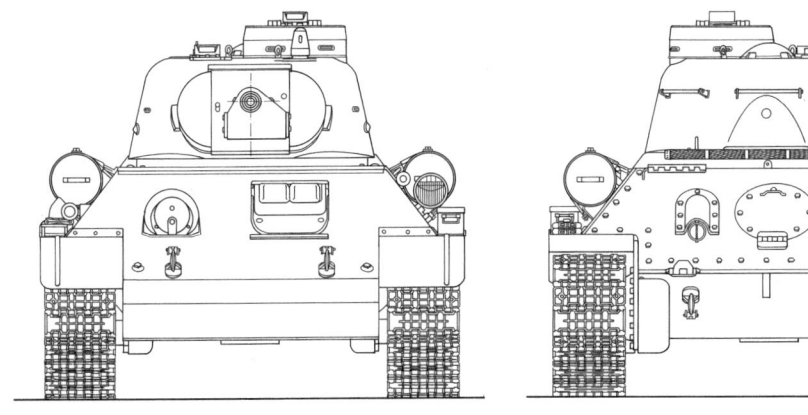

Photo showing detail of turret with commander's cupola. Missing here are the gun shield, periscopes and gunner's hatch.

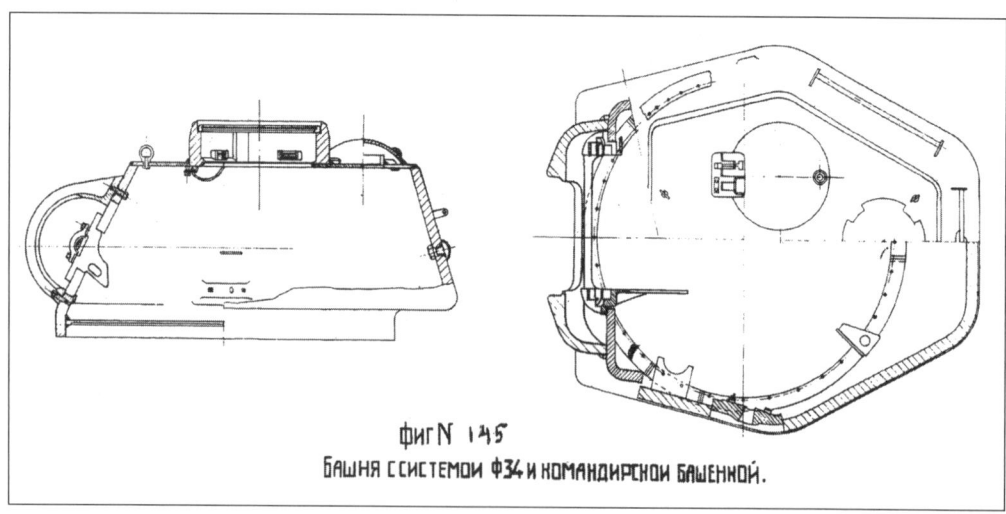

фиг N 145
БАШНЯ С СИСТЕМОЙ Ф34 И КОМАНДИРСКОЙ БАШЕННОЙ.

The turret with commander's cupola, from the batch delivered by Factory 112 Gorky, autumn 1943. This was fitted to the T-34/76 until production terminated in 1944.

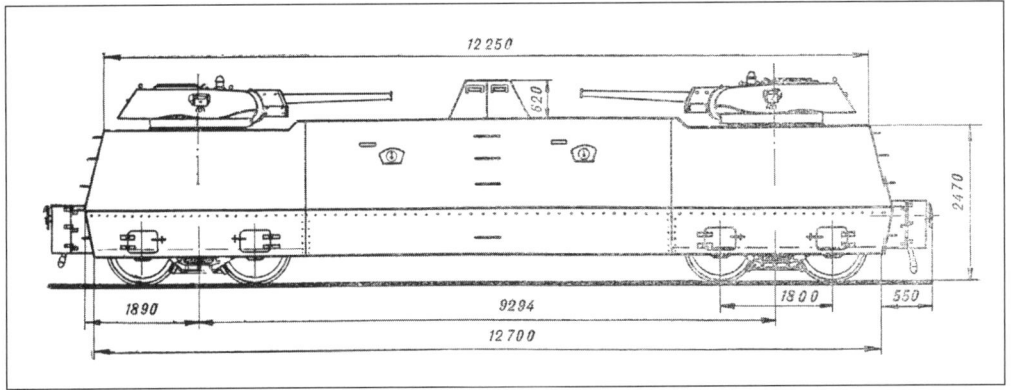

Above: T-34 turrets from various production series were frequently fitted on Soviet armoured trains. The first of this kind was armoured train No. 1 *Sa Stalina* put into service in September 1941. Its had two turrets equipped with 76.2-mm F-34 L/41.5 guns.

Right: T-34 turrets were also coveted by the Germans for use on armoured trains. In this October 1943 photograph, the train *Michael* is seen. Other T-34 turrets were installed in fortifications.

Right: Soviet river gunboats were also fitted with T-34 turrets of various series. Displacing 42–45 tonnes, length 25 m; with a turret fore and aft and a crew of seventeen, the draught was about 1m. Boats of this kind were used by river and inland waterway flotillas until the end of the war.

Hull

The hull was divided into four compartments: driver, fighting, engine and drive. It consisted of pressed steel armour plates sloping fore and aft, vertical at the sides. The superstructure was sloped all round. The roof armour covered the fighting compartment with its 1.42-m turret ring forward, the engine compartment in the centre and then the drive compartment. The cover over the engine compartment was in three parts and included a central bulge with a maintenance hatch, equipped on both sides with a lockable louvre for air intake. The armour cover for the drive

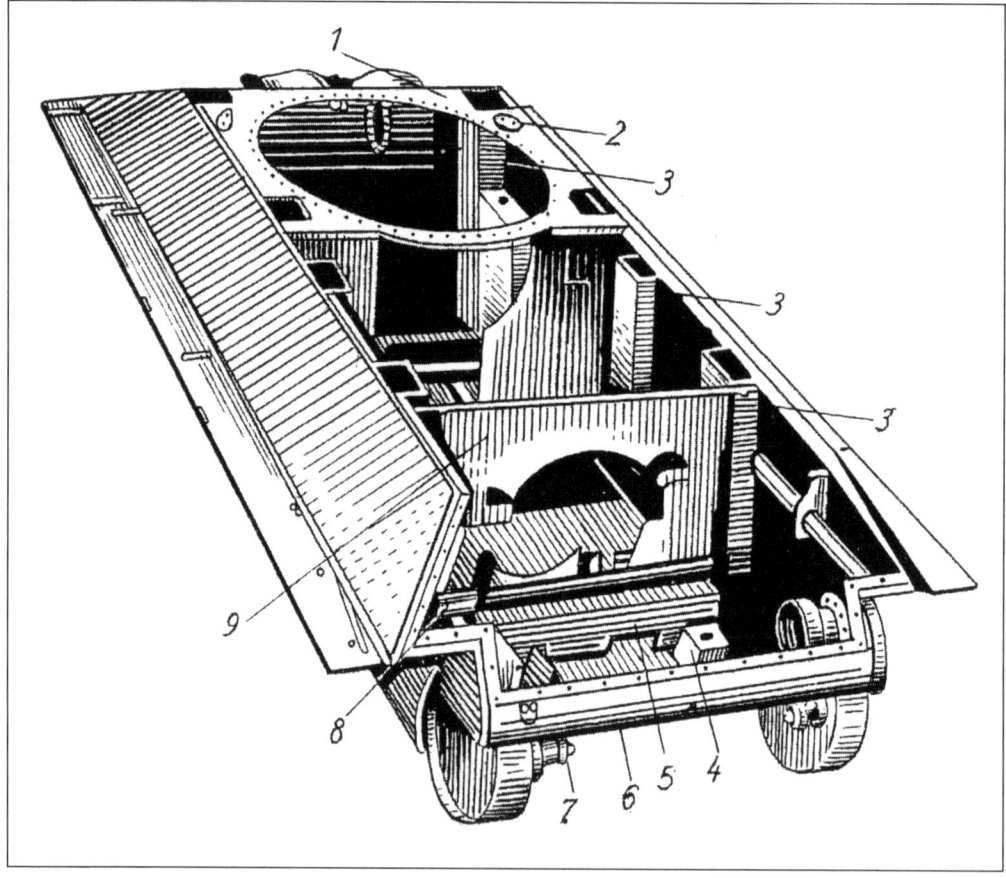

T-34 hull, internal view from rear:
Key: 1. Roof of fighting compartment; 2. Refuelling cap; 3. Spring shaft; 4. Bearing support for brake; 5. Partition floor for mounting transmission unit; 6. Lower rear armour; 7. Towing loop/eye; 8. Radial bearing; 9. Partition between engine and drive compartments (ventilator partition).

This photograph taken in July 1941 shows a T-34 manufactured as from April 1941 at Factory No. 183 Kharkov. Typical of the hull are the driver's hatch with three periscopes, the MG shield without armour plate and two bolted towing lugs. The upper and lower frontal armour is joined by a so-called cross-beam.

compartment had a hinged louvre of wire netting for ventilation. This louvre could be closed from inside by means of two hinged plates which helped regulate the air flowing out. Both the covers for the engine and the gears were attached by bolts and cap nuts to the partition walls in the interior and the lateral armour plating, and could be detached individually or completely for maintenance purposes. The rear armour plate could be raised over two hinges once the retaining bolts were released. At the rear were the two armoured hoods protecting the exhausts, and between them the maintenance hatch. This had a square form with two hinges and was bolted down. Not until 1942 did the manufacturers turn out a round maintenance hatch with a single hinge.

The upper front armour had an oval opening to the right with welded armour protection for the 7.62-mm DT MG with a telescopic sight. From 1942 this was protected by moveable armour. On the left side of the hull was the square opening with hatch cover and mechanism for the driver's

74

Left & below left: Differences in the form of the rear armour plating. The top photo shows a tank from the 1941 series, built by Factory No. 183 Kharkov, and the lower image a 1942 example from Factory No. 112 in Gorky.

hatch. The A-20 had had this solution but with only limited observation for the driver. It had been criticised because the hatch for the A-32, the following prototypes and early series vehicles had an armoured attachment with three periscopes, two of which improved the driver's vision to the right and left. During 1942 it was decided to scrap these in favour of equipping the driver's hatch with two periscopes with hinged armoured shutters. To the right of the driver was an emergency exit hatch in the hull floor below the MG gunner's position.

As their attention to the tank turret shows, designers and engineers were constantly searching for ways to improve the quality of tank hulls and rationalise manufacturing techniques. Following the receipt of plans from Factory 174 they interlocked the front, side and roof armour which increased the stiffness and resistance to shell hits. From the winter of 1942 the armoured plates were welded together automatically with flux. In 1941 the Germans had set about increasing the penetrative power of their panzer and anti-tank guns. This was achieved initially by the use of armour-piercing ammunition of increased efficiency; by using tungsten-cored and hollow-charge rounds; and by the introduction of large-calibre high-performance guns (7.5-cm and 8.8-cm).

Soviet tank builders reacted swiftly, and by the end of 1941 they had strengthened tank hulls of the current production line with welded plate 15 mm thick. In February 1942 Factory No. 112 Krasnoye Sormovo at Gorky delivered T-34s with eight-part supplementary armour; the factories at Nizhny Tagil, Stalingrad and Chelyabinsk used various versions of one- or two-part armour plate. Furthermore T-34s in repair yards received reinforced frontal armour, for example at Factory No. 28 in besieged Leningrad. These changes were responsible for a weight increase of between one and two tonnes with a corresponding loss of mobility in exchange for improved protection. Thus the 5-cm L/60 gun equipping panzers and anti-tank units which fired the hard core Pz.Gr. 40 shell found that it had no effect even at the 'suicidal close combat' range of 100 m.

Another threat came from the increasing use of high explosive armour-piercing ammunition which made it possible successfully to integrate

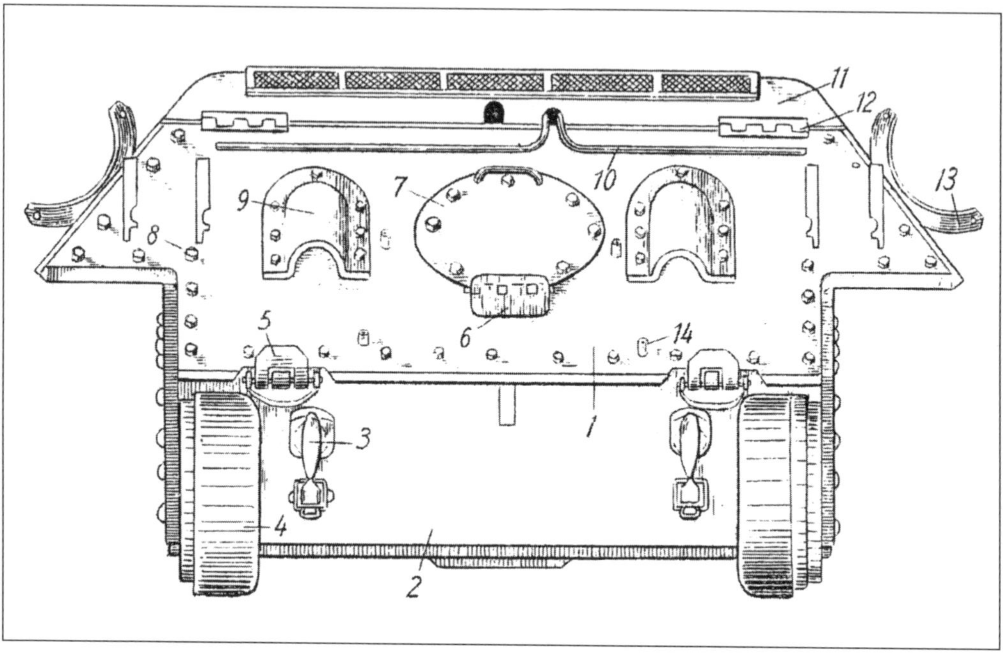

Rear view of the T-34:

Key: 1. Upper armour; 2. Lower armour; 3. Towing hook; 4. Housing for lateral gears; 5. Hinge of upper armour; 6. Hinge for hatch cover; 7. Transmission hatch; 8. Bolts for upper armour plate; 9. Armoured hood for exhaust pipe; 10. Conduit for smoke generator electrics; 11. Transmission housing grille; 12. Wire grille; 13. Fuel container supports; 14. Socket for pre-heating equipment.

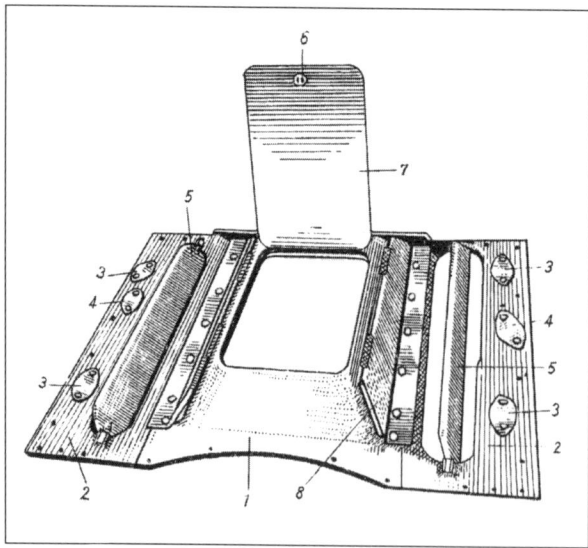

Cover for engine compartment:
Key: 1. Central plate; 2. Lateral plates; 3. Armoured cover above spring shafts; 4. Refuelling cap; 5. Louvre plates over the hot air extraction vents; 6. Cover lock; 7. Hatch cover; 8. Steel last.

Photograph taken during fighting in the northern section of the Eastern Front, early 1943, showing a T-34s engine compartment cover torn away.

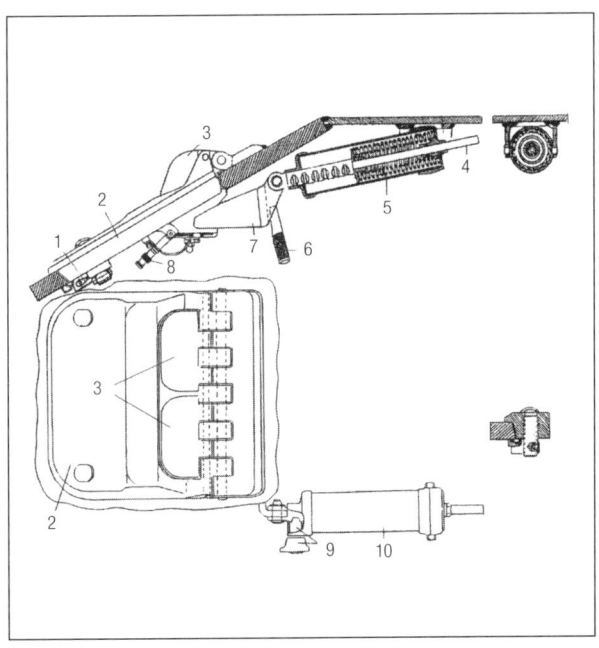

Overall and detail views of the driver's hatch and its operating mechanism:
Key: 1. Snap catch; 2. Armoured hatch cover; 3. Periscope shield; 4. Hinge rod; 5. Hydraulic hinge mechanism; 6. Handle for driver's hatch; 7. Lever arm of hatch cover; 8. Grip for periscope shield; 9. Locking device and hand-wheel for adjusting gear; 10. Housing for adjusting gear.

A 1941 model T-34 (*left*) and 1942 model (*nearer camera*) originating from the Stalingrad Tractor Factory. The rear tank (with 3-kg limpet mine attached below the driver's hatch) has armour plate welded on the turret side – likely a repair of battle damage. The tank in the foreground has the eight-part supplementary armour on the hull front and the reinforced turret armour.

1942-built T-34 from Factory No. 112 Gorky. Notice the MG mount with armoured plate. The driver's hatch is reinforced to 72 mm with protective cover and has two periscopes.

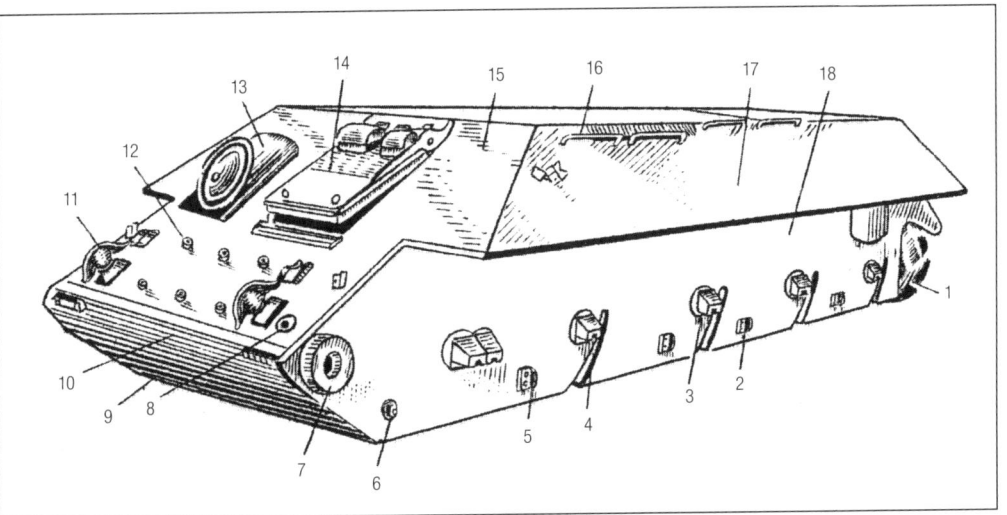

Hull (exterior view):
Key: 1. Rear gear and sprocket mounting; 2. Mounting for axle suspension sprockets; 3. Suspension sprocket limiter or stop; 4. Vertical coil spring housing; 5. Mounting for sprung suspension sprockets; 6. Mounting for front suspension linkage; 7. Idler mounting; 8. Locking screws for track tension device; 9. Lower glacis; 10. Glacis armour plate; 11. Tow eyelets; 12. Spare track shoe brackets; 13. Armoured mount for hull machine gun; 14. Driver's hatch; 15. Upper glacis plate; 16. Hand rails for tank riders; 17. Sloping side armour; 18. Lower hull side plate.

---

guns with low muzzle velocity – infantry guns and howitzers – into anti-tank defences. The use of hollow-charge ammunition determined the penetrative effect of an increasing number and variety of close-range anti-tank weapons. The Soviets made several attempts to counter this threat. From Factory No. 112 Krasnoye Sormovo came a suggestion for strengthening the hull and turret of the T-34 by the use of armour skirts. These were not introduced officially: the troops devised their own as they saw fit, mainly later in the war as a means to protect the tank against close-range Panzerfaust rockets that could penetrate 200 mm of armour.

T-34 hulls had minor differences in towing hooks and the arrangement of boxes for tools and crew baggage, as well as the number and choice of supplementary fuel tanks – as a rule these were round barrels for 135 litres of diesel. At the end of 1942, Factories No. 112 and No. 183 delivered T-34s with squared containers at the rear. At about the same time additional handholds began to appear on tank hulls for infantry tank riders.

Above: 1942 model T-34, with reinforced hull front and turret armour. Photo taken in 1942 during the siege of Leningrad.

Right: Square extra tanks at the rear of a captured T-34 originating from Factory No. 112 Gorky. Photo taken in the Kursk salient, July 1943.

Fuel in barrels, two per side, was frequently found. This T-34 from Factory No. 183 Nizhny Tagil was put out of action by hits in the fuel tanks from a German 2-cm Flak 38.

Armament

In January 1940 the Defence Committee at the Council of the People's Commissariat decided to adopt the 76.2-mm F-32 L/31 gun. It had been tested the previous year in the BT-7, but for the current series production of the T-34 and KV was not available in sufficient numbers. In order to close the gaps in the series run, the reserve variant 76.2-mm L-11 L/31.5 had to be fitted. It was only slightly heavier and reasonably efficient but inferior to the F-32. However, several hundred of these guns from the 1938–9 run were ready for immediate fitting.

Work was continuing meanwhile on the development of the 76.2-mm F-32 L/31 under the direction of chief designer V. G. Grabin. The result was the improved F-34 L/41.5 gun, the first example being ready in September 1940. It was convincing in every respect. Though only slightly heavier it had greater range (maximum 11,200 m), greater penetration

1941 model T-34, manufacturer Factory No. 183 Kharkov, on display at the Kubinka Tank Museum near Moscow.

The hybrid BT-2 was a development of the Christie M.1931 tank armed with a 37-mm L/37 gun and a 7.62-mm MG: armour was 6–13 mm. During 1932–3, the Red Army received 620 BT-2s and well over 4,000 of its successors, the BT-5 and the BT-7.

Fragment of a T-34 hull from the 1941 series (driver's hatch with three periscopes). Parts of the additional armour are missing. The retaining hooks for the spare track segments can be clearly seen here.

1943 model T-34 with 76.2-mm F-34 L/41.5 gun with commander's cupola fitted, manufactured from April 1943 at Uralmash, Sverdlovsk.

1942 model T-34 with 76.2-mm F-34 L/41.5 gun in an improved turret, as delivered from Nizhny Tagil from April 1943.

Rear armour plating, as built in a 1941 series T-34 from Factory 183, Kharkov.

Tracks of the later form. A complete track of seventy-two segments weighed 1.07 tonnes.

Drive sprocket of a later production run. Of cast steel, it had a diameter of 50 cm and together with the track tension device weighed 220 kg.

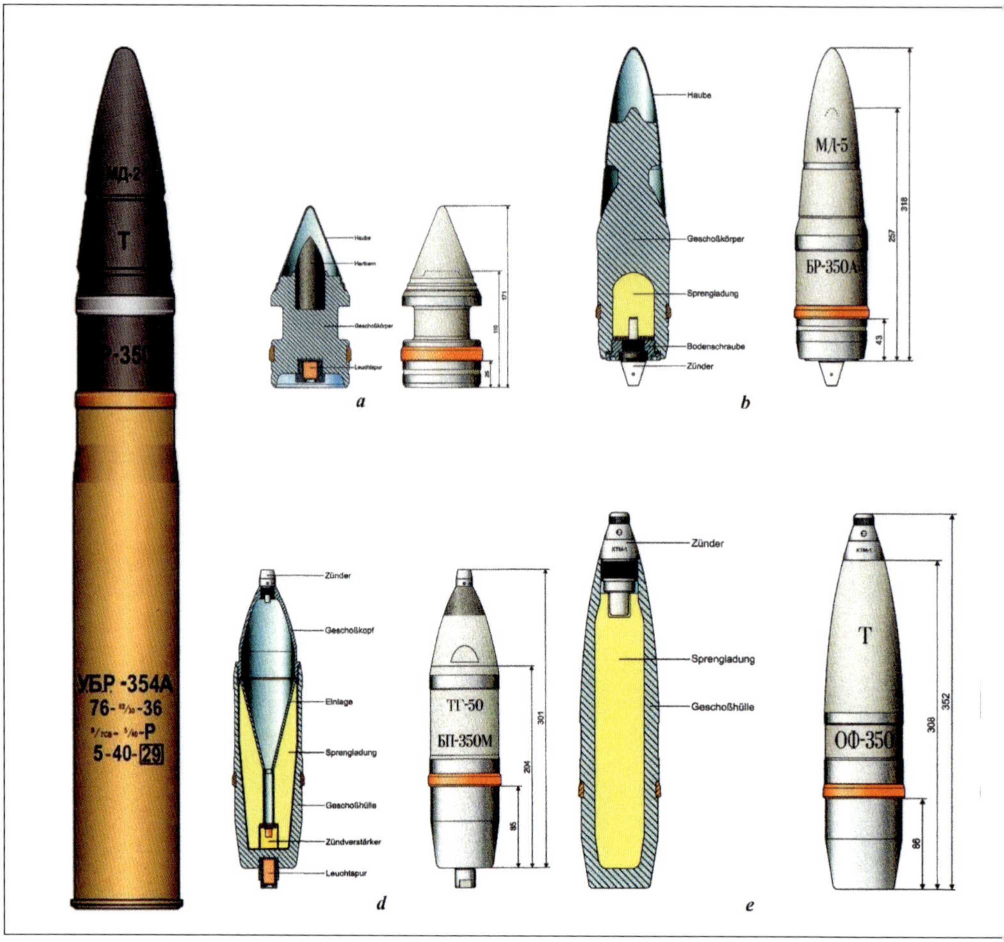

The 76.2-mm F-34 L/41.5 fired fixed ammunition with an overall length of 61 cm for the UBR-354A AP round: (a) 76.2-mm BR-354P AP (sub-calibre); (b) 76.2-mm BR-350A AP; (c) 76.2-mm BR-350B AP; (d) 76.2-mm BR-350M AP (hollow charge); (e) 76.2-mm OF-350AP explosive-fragmentation; (f) 76.2-mm D-350A smoke.

Views of two gun mantlets from below, showing the rough workmanship in mass production by the individual manufacturers.

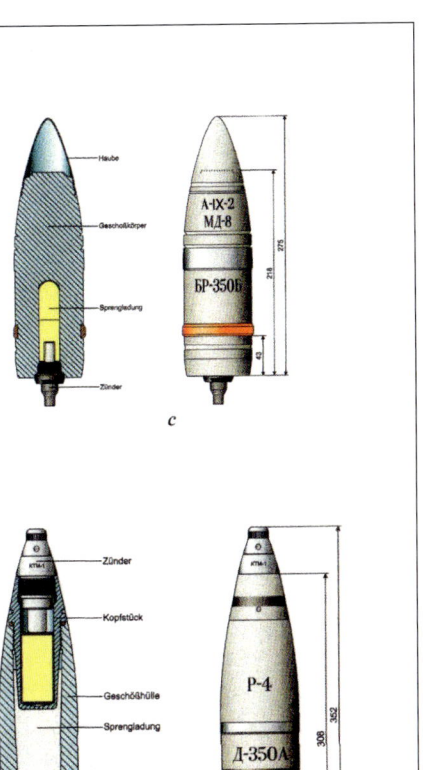

Top right: Towing hook from a model 1944 T-34/85.

Above right: External view of the hull 7.62-mm DT MG in the same tank.

Right: Front view of T-34 with cast turret. The shield for the hull MG is missing.

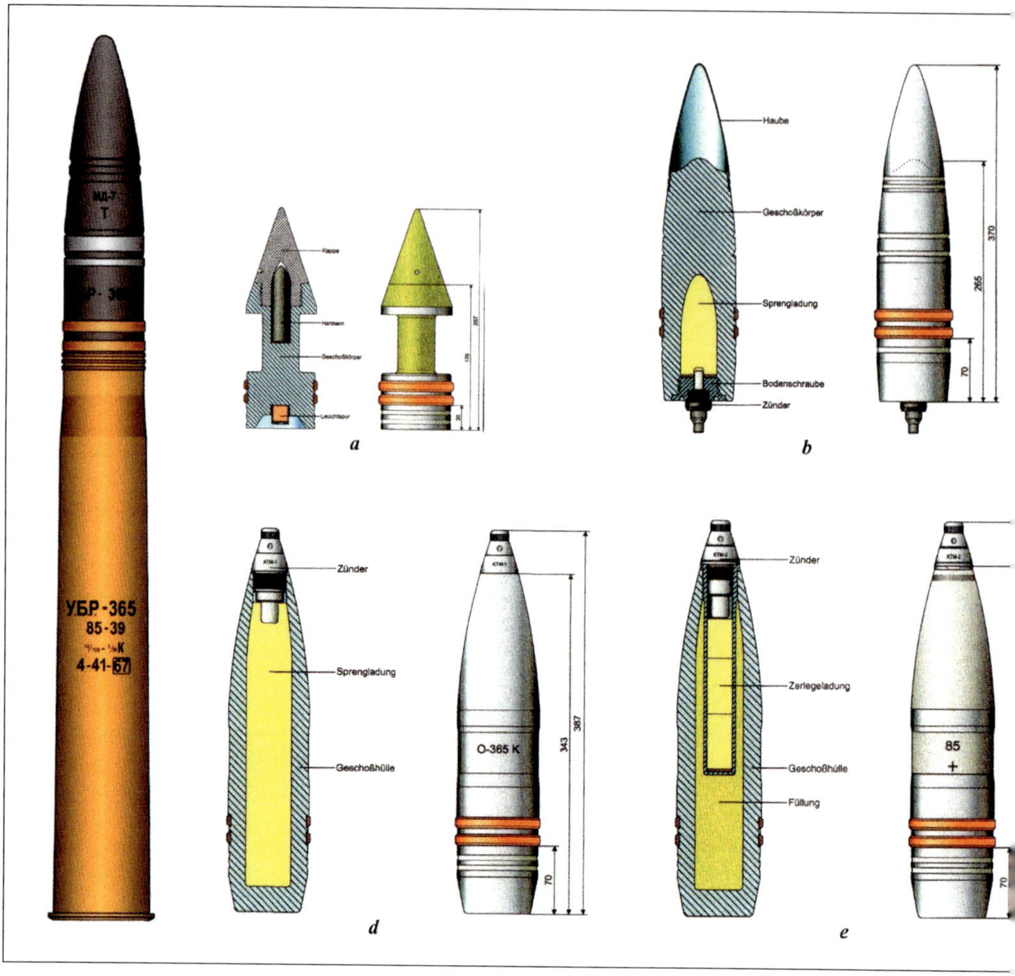
МД-7
Т
УБР-365
85-39
4-41-67
Kappe
Hartkern
Geschoßkörper
Leuchtspur
a
Haube
Geschoßkörper
Sprengladung
Bodenschraube
Zünder
370
265
70
b
Zünder
Sprengladung
Geschoßhülle
О-365 К
343
387
70
d
Zünder
Zerlegeladung
Geschoßhülle
Füllung
85
+
70
e

327

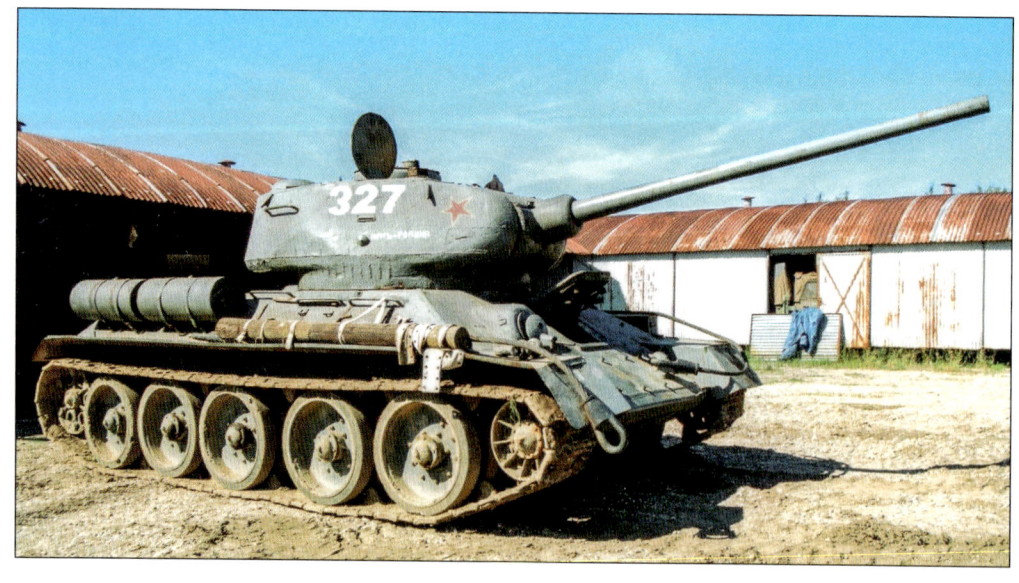

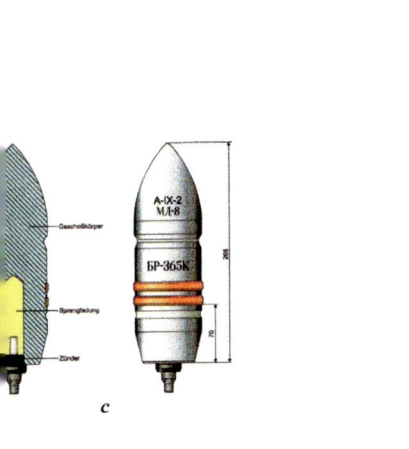

The 85-mm ZIS-S-53 L/54.6 fired fixed ammunition with an overall length of 920 mm for the UBR 365 AP round:
(a) 85-mm BR-365P AP (sub-calibre);
(b) 85-mm BR 365 AP;
(c) 85-mm BR-365K AP;
(d) 85-mm O-365-K fragmentation;
(e) 85-mm D-367 smoke.

Below left: A 1944 model T-34/85 fitted with an 85-mm ZIS-S-53 L/54.6 from Factory No. 183 Nizhny Tagil, delivered from May 1944.

Below: The T-34/85 was the Red Army's most important tank in 1944–5 and fought in all European theatres and in the Far East against Japan.

A T-44A with the 85-mm ZIS-S-53 L/54.6 gun. In 1944–5, 655 tanks of this type were delivered. With a four-man crew, its weight was 31.5 tonnes and it had a top speed of 55 km/h.

Completed T-34/85 from Factory No. 183 Nizhny Tagil.

There were minor differences between T-34/85s from different manufacturers. Above and below (both) are examples from Nizhny Tagil.

Right: A T-34/85 from Factory No. 112 Gorky.

An SU-122 gun at the Kubinka Tank Museum.

Scrapyard finds:
(*above*) a 100-mm gun with damaged barrel. With barrel, breech and mechanism it weighed 1,435 kg;
(*right*) a rather battered SU-100 hull, but with many of its original features still intact.

Above: An SU-100 from the Urals Machinery Factory, Sverdlovsk, seen at the Kubinka Tank Museum, near Moscow.

Rear view of an SU-100 (maintenance hatch missing).

Detailed view of gun shield and driver's hatch with two periscopes, armoured covers lowered. Above them is the pistol port for the 7.62-mm PPSh SMG.

Armoured covers for the two ventilators on the right side of the vehicle.

SU-100 commander's cupola with base for aerial.

All three photos: A selection of SU-100s from the Uralmash factory, Sverdlovsk, showing minor construction differences:
(*above*) from the collection at MM Park France, Wantzenau (near Strasbourg);
(*top left*) German-Russian Museum, Berlin-Karlshorst;
(*left*) Military Collection at Senica (near Bratislava).

The wreck of an SU-100 with engine and drive unit removed. Noteworthy is the arrangement of fuel tanks and notice the pistol port in the rear wall of the casemate superstructure. Photo taken at 'Tank Farm Nasielsk'.

The SU-101 of 1945 marked the end of the development of self-propelled guns on the T-34 chassis and fitted with the 100-mm D-5-S L/56.

The 76.2-mm L-11 L/30.5 gun. Notice the cast gun housing, ventilator in the roof armour (on the left in the photo), the PT-6 periscope (× 2.5 and 26°) and the all-round observation equipment in the open turret hatch.

Production of L-11 and F-34 76.2-mm Guns

Year	76.2 mm L-11 L/30.5	76.2 mm F-34 L/41.5
1938	570	–
1939	176	–
1940	–	50
1941	–	3,470
1942	–	14,307
1943	–	17,161
1944	–	3,592
1945	–	–
Totals	746	38,580

The gun in the turret of a T-34 manufactured at the Stalingrad Tractor Factory in 1941 (total production for the year was 956 units). Also seen by the turret hatch lid is the signal flag hatch.

(70 mm at 500 m) and a shorter recoil (320–390 mm as against the 450 mm of the other models, the latter being an important advantage given the limited space in the T-34). The F-34 had a semi-automatic wedge-type breech block, a ratchet-type elevating mechanism, hydraulic barrel brake and hydraulic-pneumatic recoil mechanism. The TOD-7 Model 1941 ($2.5 \times 15°$) telescopic sight and the $2.5 \times 26°$ periscope-telescope replaced the former items. The new gun was installed into a T-34 for the first time in November 1940 and tested on the Gorochovetz firing range. As a result, the decision was taken to use the F-34 as the gun of choice in the tank. The first series-produced T-34 armed with it left the Kharkov factory in March 1941.

Two 7.62-mm DT MGs (Degtyarev system), one of them co-axial to the right near the gun, the other in the right of the hull, completed the armament. The air-cooled gas-pressure loader was installed for firing 7.62-mm × 54R rounds. The ammunition was supplied from 47-round drum magazines.

The 76.2-mm F-34 L/41.5 gun fired fixed ammunition. The AP shells with base detonator (amongst others the BR b350A type) weighed between 6.3 and 6.51 kg, fragmentation shells with nose detonator (OF 360 and O.350A) between 6.2 and 6.4 kg. During the war smaller calibre hard-core shells (BR 350P) were introduced which weighed only 3.02 kg and could penetrate armour between 77 mm and 92 mm thick at 500 m. At a greater range the penetration fell off sharply in contrast to the heavy hollow-charge shells (BP 353A etc.) weighing from 3.94 kg to 5.28 kg able to penetrate 70 mm of armour regardless of range. Despite all efforts, the possibility of improving the efficiency of 76.2-mm tank guns remained limited

The design office of Leningrad Factory No. 92 had been looking at a 57-mm calibre gun in 1940, and this resulted in the 57-mm F-31 anti-tank gun, the AP shells of which, fired at a muzzle velocity of about 1,000 m/sec, could penetrate 70 mm of armour at 500 m. In 1941, 371 guns designated 57-mm ZIS-2 were completed. Because of problems in production and the lack of need for such performance in action at the time, the future orders were cancelled and production was not resumed until 1943.

In parallel with the anti-tank gun, in the spring of 1941 a 57-mm gun had been required for the T-34. It was fitted experimentally to a T-34 of the 1941 series and tried out the same month at the Sofrino firing range. It was found to offer an efficient alternative to the 76.2-mm F-34 L/41.5 gun, supplies of which at this time were lagging far behind requirements. The gun received the designation 57-mm ZIS-4 L/73. The TMFD (2.2 × 15°) telescopic sight was fitted. It has been suggested by various writers that forty T-34s were fitted with this gun in the autumn of 1941, but the only proof so far is that the 21st Tank Brigade formed at Vladimir on 10 October

1941 model T-34 with 57-mm ZIS-4 L/73 gun during testing. With a four-man crew, the tank weighed 26.8 tonnes. In the summer of 1943, Factory No. 183 turned out a new 57-mm gun (ZIS-4M L/73), with which the 1942 model T-34 weighed 29 tonnes.

Performance of Soviet Tank Guns

	45-mm M.1934 L/46	45-mm M.1938 L/46	57-mm ZIS-4 L/73	76.2-mm L-10 L/23.7	76.2-mm L-11 L/30.5
Calibre (mm)	45	45	57	76.2	76.2
Muzzle velocity AP/HE shells (m/sec)	760/335	760/335	995	555/560	612/635
Rate of fire (rpm)	12	10–12	6–10	6–7	6–7
Barrel length (m)	2.09 = L/46	2.09 = L/46	4.16 = L/73	1.801 = L/23.7	2.32 = L/30
Length of inner lining (m)	1.975	1.975	–	–	2.188
Recoil brake	Liquid	Liquid	Liquid	Liquid	Liquid
Recuperator	Spring	Spring	–	Pneumatic	Pneumatic
Method of firing	Mechanical	Mechanical and electric	Mechanical	Mechanical	Mechanical
Weight, barrel and breech (kg)	135.5	135.5	–	433	437
Weight, ready to fire (kg)	–	–	–	–	–
Range, effective (m)	c. 900	c. 900	c. 1,100	c. 700	c. 750
Range, maximum (m)	4,400	4,400	12,500	5,600	5,600
Armour penetration at 500 m (mm)	31/35	31/35	89/105	40[3]	50/66[3]
Tank types used on	BT-7, BT-8, T-26, T-46-5	A-20, A-32, T-26	T-34 Model 1941, 1942 1943	A-32	T-34 Model 1940, 194

NOTES: 1. Sub-calibre shell 950 m/sec; 2. Sub-calibre shell 1,154 m/sec; 3. At 1,000 m.

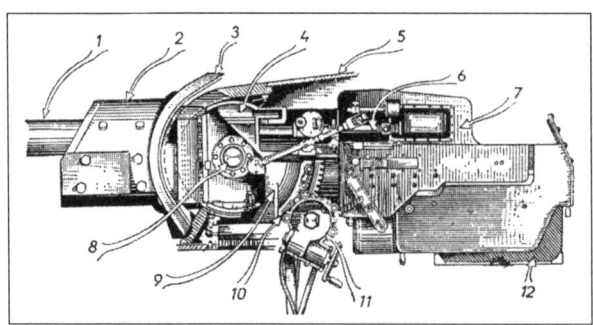

Side view of the F-34 L/41.5 gun:
Key: 1. Barrel; 2. Armour protection; 3. Mantlet; 4. Sighting device; 5. Turret; 6. Optic; 7. Deflector shield; 8. Pivot pin; 9. Barrel cradle; 10. Toothed segment; 11. Elevating gear; 12. Cartridge case collector.

6.2-mm F-32 L/31	76.2-mm F-34 L/41.5 (ZIS-5)	76.2-mm S-54 L/58	85-mm D-5-T L/51.6	85-mm ZIS-S-53 L54.6	100-mm D-10-K L/53.5
76.2	76.2	76.2	85	8	100
613/638	662/680[1]	816/800[2]	792/785	792/785	880
6–7	5	3–5	5–8	6–10	4–6
.362 = L/31	3.17 = L/41.5	4.42 = L/58	4.386 = L/51.6	4.64 = L/54.6	5.348 = L/53.5
–	2.962	–	–	3.495	–
Liquid	Liquid	Liquid	Liquid	Liquid	Liquid
Pneumatic	Pneumatic	Pneumatic	Pneumatic	Pneumatic	Pneumatic
Mechanical	Mechanical	Mechanical	Mechanical	Mechanical	–
434	538	–	980	905	1,538
–	1,155	1,390	1,500	1,150	2,257
c. 750	800–900	c. 950	900–950	900–950	1,000 m
10,000	11,200	13,000	12,900	13,600	16,000
70	77/92	–	90/140	90/140	125/155
KV-1, A-34	T-34 Model 1941 bis 1943, A-43 (Project)	T-34 Model 1943	T-34/85	T-34/85, T-44/85	T-44/100

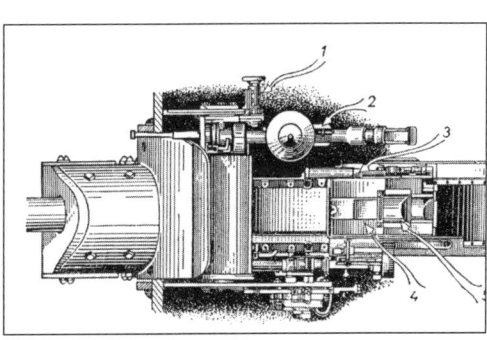

Overhead view of the F-34 gun:
Key: 1. Gun lashing; 2. Coaxial MG; 3. Semi-automatic.; 4. Base plate; 5. Wedge lock.

View into the turret of a captured T-34. Left, the commander/gunner's position, to the right the loader's position, between them the breech of the 76.2-mm F-34 L/41.5.

1941 was equipped with BT-7 tanks, a light T-60 and several T-34s with standard armament plus ten armed with the 57-mm ZIS-4 L/73. These were all lost in the heavy fighting around Kalinin in the second half of October 1941.

Another tank with the 57-mm ZIS-4 L/73 gun, the T-44 in the autumn of the year, remained only a project. Later the 57-mm calibre played a part in T-34 armament in the summer of 1943. Under the pressure of German successes with the Tiger tank, a 57-mm gun was called for again, and the 57-mm ZIS-4, now with higher-performance AP ammunition able to penetrate between 83 mm and 105 mm armour, should have been enough to deploy the T-34 successfully as a 'panzer destroyer'. A tank of the current production received this gun, and was trialled between May and August 1943 but it never entered series production.

In parallel with the 57-mm gun, two other methods were tried in an effort to increase the chances of winning through in the struggle

Training model of the 76.2-mm F-34 L/41.5 gun.

The 76.2-mm F-34 L/41.5 in a hexagonal turret, manufactured at the Uralmash Factory, Sverdlovsk. In the background is a turret with commander's cupola usual on the T-34/76 version (without the mantlet and with only fragments of the gun shroud).

Overhead view of the T-34 turret with 76.2-mm F-34 L/41.5 gun as turned out in 1942 by Factory No. 183 Nizhny Tagil.

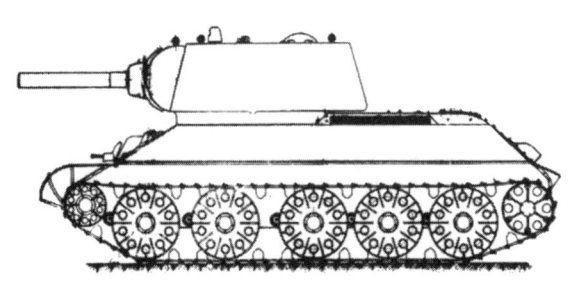
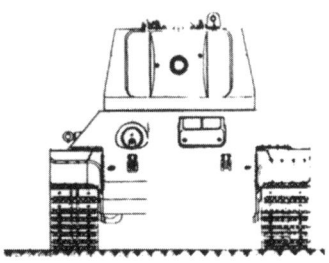
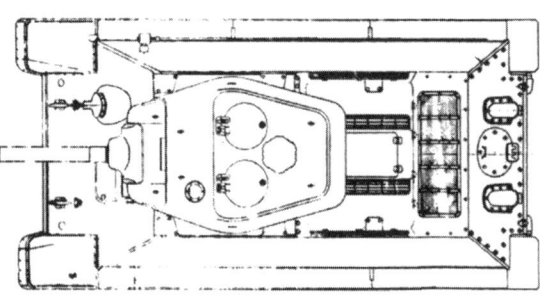

The project for a T-34 with the 122-mm U-11 L/22.7 howitzer in a version from Factory No. 8 Sverdlovsk, May 1942.

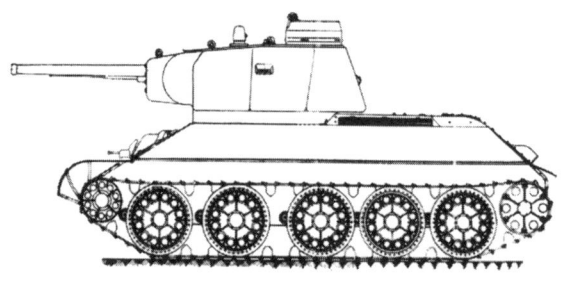
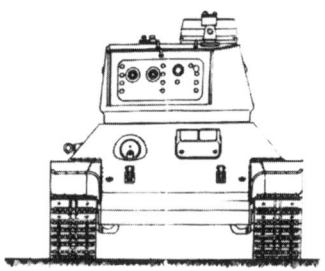
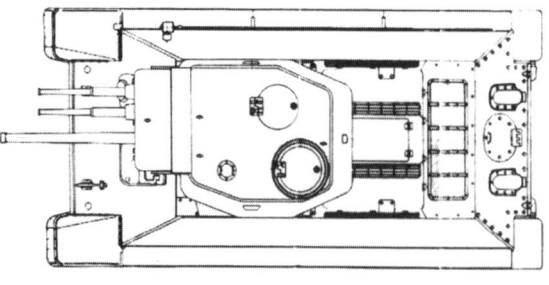

The T-34 project as an artillery support tank with a 76.2-mm gun and two 45-mm guns (Factory No. 183, Nizhny Tagil, 1942).

against the German panzers. The first was to lengthen the barrel of the 76.2-mm gun from the calibre length of L/41.5 to L/58. This gun was designated the S-54. Although shells could be fired at a muzzle velocity of 816 m/sec − and the 3.05-kg sub-calibre shells even at 1,154 m/sec − apparently these did not meet expectations. At least one T-34, built in 1942, was tried out in October 1943 with the S-54 (elevation -4.5° to +26.5°). Sixty-eight main-gun rounds and 2,331 MG rounds made up the ammunition load. A requirement to increase the penetration to 100 mm (from 80 mm) shows the limit of the calibre.

The other method was to increase the calibre of the main weapon of medium tanks to 85-mm with a prospect of 100-mm later. The installation and trials of an 85-mm S-53 L/54.6 in a 1943-built T34/76 soon showed

1943 model T-34 fitted with the 85-mm S-53 L/54.6 gun. The tank weighed 32.2 tonnes with a five-man crew. The gun could not be loaded or fired while the tank was on the move.

Model 1940/1941 T-34, from Factory No. 183 Kharkov. All road wheels have hard rubber tyres. This tank was captured by the Slovak Army in the summer of 1941.

very clearly that on account of lack of space, a new design of turret was unavoidable. The development was the T-43 and T-34/85.

Other considerations for greater T-34 firepower got no further than the drawing board, and nor did an artillery tank with larger turret, and a 122-mm U-11 L/22.7 howitzer. Another unusual idea was the project for a T-34 with one 76.2-mm and two 45-mm guns. This was worked on at the Nizhny Tagil design bureau of Factory No. 183. The outward features included a heptagonal turret with a broad mantlet, and a high cupola for the commander.

Road Wheels and Tracks

The road wheels had hard rubber tyres and consisted of two cast, later press-stamped, disc wheels and a hub. They were connected to the coil springs by rocker arms which allowed a tolerance of 240 mm. This very yielding suspension system provided good driving performance. The rubber tyres reduced noise and increased track stability while the tyres themselves had a cushioning effect which helped extend the working life of the wheels.

In 1942, several manufacturers began making lighter press-stamped disc wheels with holes in the disc and spokes, but continued with the typical hard rubber tyres. However, the shortage of rubber in the winter of 1941 forced the introduction of pierced steel wheels with rubber rings. This measure replaced 90 per cent of the hard rubber for a weight increase of 5 per cent per road wheel. A substantially shorter working life of the wheel body and tracks had to be allowed for. Some manufacturers compromised by fitting wheels with rubber tyres at the front and rear, and three steel wheels between them. Up until the beginning of 1943, the Stalingrad Tractor Works supplied the T-34 with almost exclusively steel wheels. Once the supply of rubber was guaranteed, wheels with hard-rubber tyres dominated once more.

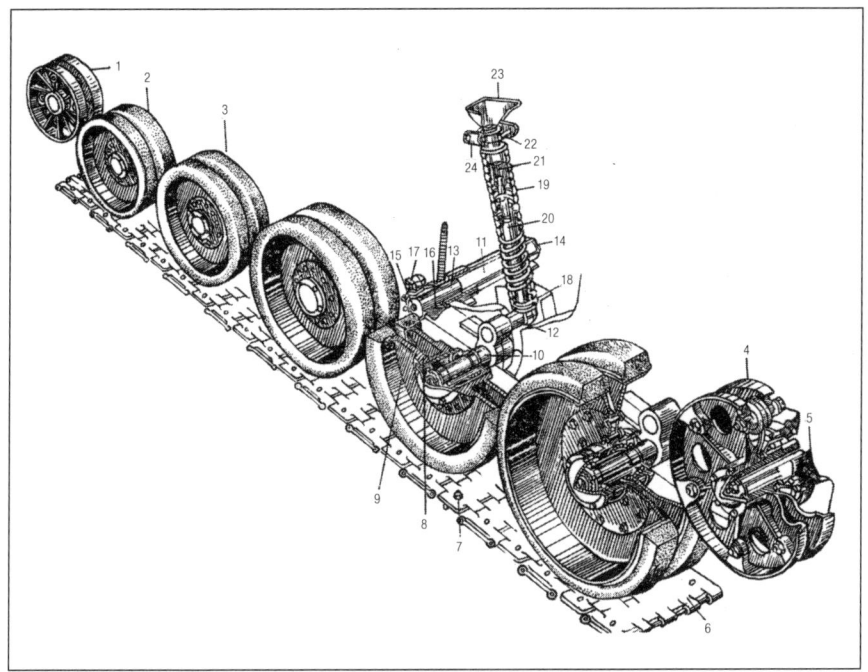

Running gear of the T-34/76 (model 1943) and T-34/85:
Key: 1. Idler wheel; 2 & 3. Road wheels; 4. Drive sprocket; 5. Drive shaft; 6. Track shoe; 7. Securing bolt for spur; 8. Hub; 9. Steel road wheel with rubber tyre; 10. Road wheel axle; 11. Swing-arm axis; 12. Spigot/pivot; 13 & 14. Sleeves; 15. Locking nut; 16. Suspension sprocket; 17. Suspension sprocket limiter; 18. Disc; 19. Vertical coil springs; 20. Centralised control rod; 21. Cap for centralised control rod; 22. Spring suspension bolt; 23. Dustproof covering of suspension system; 24. Suspension spring.

Examples of Various T-34/76 Running Gear Designs

Model 1941/1942 T-34 from the Stalingrad Tractor Factory. All road wheels are of steel with rubber tyres. Photo from summer 1942.

Model 1942 T-34, from Factory No. 183 Nizhny Tagil. Three of the five road wheels on either side are of steel with rubber tyres. Photo from 1942.

T-34 as delivered from factories No. 112 Gorky and No. 183 Nizhny Tagil in 1943–4. All road wheels are machine punched and provided with rubber tyres. Photo from 1944.

Two T-34s with differing running gear. Both were manufactured at the Stalingrad Tractor Factory and attached to the 172nd Tank Regiment.

Left: Track segments added to the side of a T-34 for additional protection. Photo taken north of Lake Peipus, March 1944.

Below: A strip of spurs for adding to a track segment without projection in order to obtain a better group on terrain.

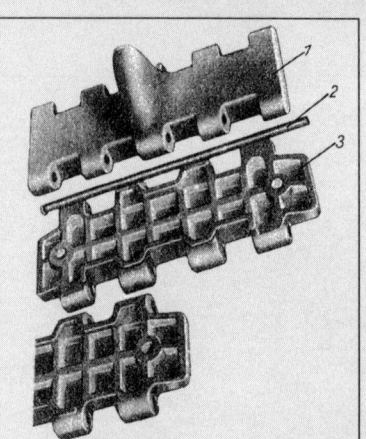

Left: 1. Track shoe with projection for grip on terrain; 2. Connecting dry pin; 3. Track shoe without grip projection.

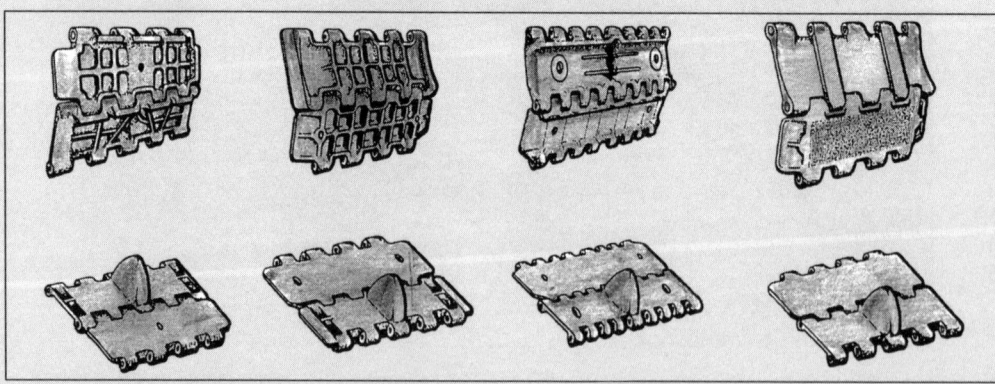

Various T-34 track shoes.

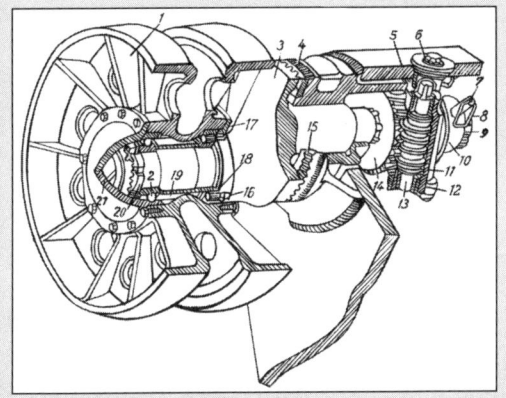 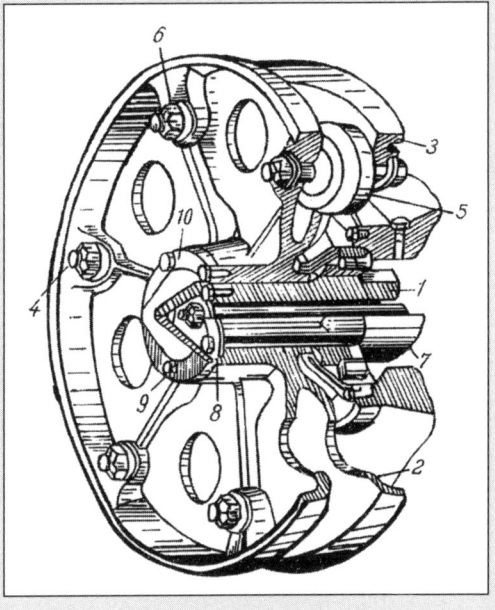

Idler wheel with track tensioner:
Key: 1. Idler wheel; 2. Ring groove bearing; 3. Eccentric; 4. Outer bearing of eccentric axle; 5. Bush/sleeve/liner/socket in drive housing; 6. Lock or catch; 7. Housing for locking device; 8. Handle to open locking device; 9. Eccentric special nut; 10. Cover; 11. Drive housing; 12. Flange of housing; 13. Worm of worm gear; 14. Worm gear; 15. Ring with Hirth coupling; 16. Sealing; 17. Cover for sealing; 18. Ring cylinder bearing; 19. Spacer bushing; 20. Castellated nut; 21. Protective cap.

Cast drive sprocket:
Key: 1. Hub; 2. Flange; 3. Rim; 4. Roller axis; 5. Transport roller; 6. Castellated nut; 7. Driven shaft of lateral countershaft; 8. Ring securing drive wheel; 9. Screw; 10. Protective cap.

Idler wheel of a T-34 from the 1942 run.

Rear view of a T-34/76, 1943/1944 production runs. Notice the cast-steel drive wheel.

T-34, complete with steel road wheels and rubber tyres (Stalingrad Tractor Factory, 1941). The turret shape and the form of the running gear illustrate clearly how the transition flowed between the individual versions.

Although the road wheels and tracks bore a resemblance to those of the BT-class tanks, on the T-34 both were broader. On either side the track had seventy-two cast manganese steel segments, each second one with an upright projection. Segments were joined with non-lubricated, unlocked track bolts passing through each segment from one side to the other. Initially the track segments were too weak and were often changed as the war went on. Cross-country performance was enhanced by a fine streamlining of the contact surfaces. In addition a strip of spurs could be screwed to all track segments lacking the protrusion (thirty-six each side). These helped increase the grip on hard surfaces, for example on frozen ground.

The drive sprockets at the rear were cast-steel double disc wheels without rubber tyres. The principal purpose of the five rolling wheels per side was to support the track while the raised leading wheel and tightening equipment at the front of the undercarriage controlled the tracks in driving manoeuvres and when turning. The sprocket and leading wheels were both subject to minor changes in the course of the war.

A T-34 from Factory No. 183 Kharkov with tracks of an earlier form. Photo taken at Orel in October 1941.

Drive Installation

The 12-cylinder 4-stroke 500-hp W-2 diesel engine in a light metal housing was comparatively light at 812 kg. It was considered reliable and low on fuel consumption. In a 1943 analysis by the German Maybach Motorenbau GmbH, the engine was assessed as 'first class', all cast-steel and mechanical sections being of remarkable quality. It was installed in the T-34 throughout the war and found use in many variant types. Until 1940 it was only manufactured by Motor Factory No. 75 in Kharkov, until this

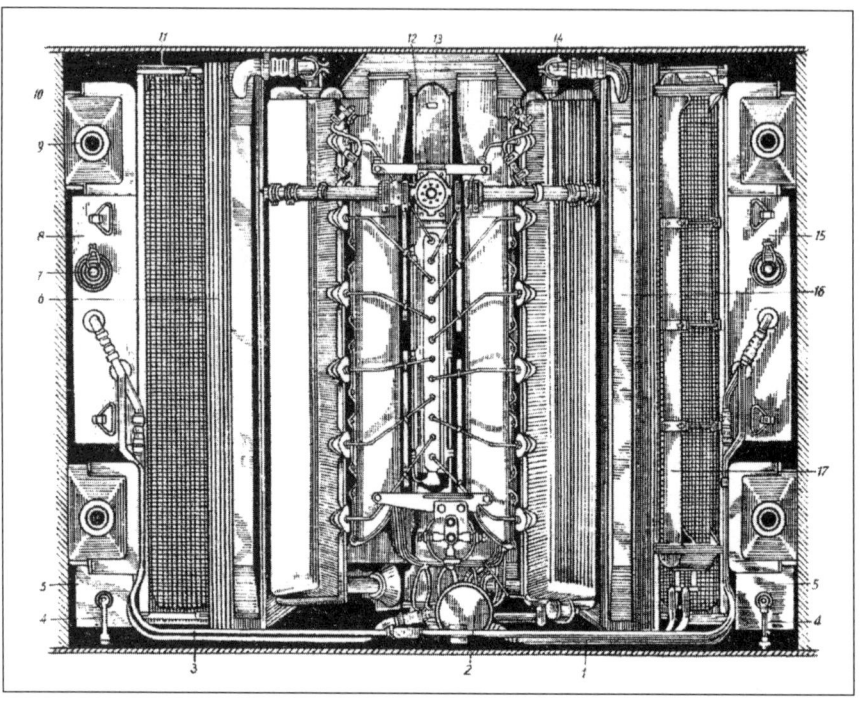

View of engine compartment from above:
Key: 1. Oil feed; 2. Oil adjuster; 3. Air pipe from oil tank to crankcase; 4. Air pipe; 5. Fuel tank; 6. Water cooler right side; 7. Locking screw for oil filler cap; 8. Oil control right side; 9. Suspension spring; 10. Shaft for spring; 11. Radiator grille; 12. Lock for water feed; 13. W-2 34 engine; 14. Water pipe; 15. Oil tank left side; 16. Water cooler left side; 17. Oil cooler.

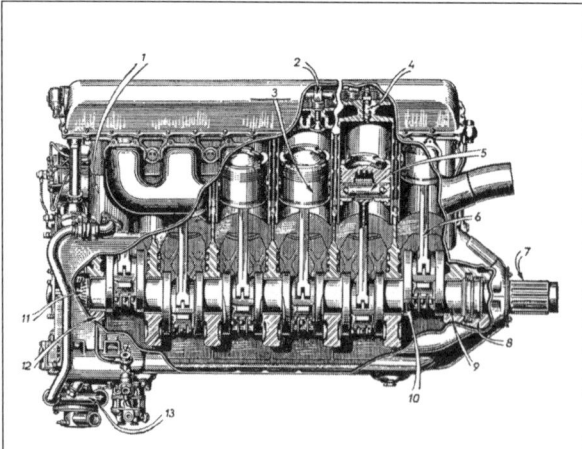

W-2 diesel engine, longitudinal section:
Key: 1. Cylinder block; 2. Camshaft; 3. Piston; 4. Injection nozzle; 5. Piston pin; 6. Connecting rod; 7. Stump of crankshaft; 8. Main bearing of crankshaft; 9. Main spigot; 10. Camshaft; 11. Camshaft drive; 12. Main bearing of crankshaft; 13. Water pump.

T-34 four-speed gearbox:
Key: 1. Lower housing; 2. Main shaft; 3. Fixed bolt; 4. Upper housing; 5. Grease nipple; 6. Piston rod; 7. Slat; 8. Gear lever for reverse; 9. Lever for 3rd and 4th gear; 10. Harness for starter relay; 11. Housing for lever for 3rd and 4th gear; 12. Cover for reverse gear cogs; 13. Air vent; 14. Bolt for change column; 15. Surface for starter bracket; 16. Sealing ring nut; 17. Spur gear; 18. Check-screw; 19. Support plate for gearbox; 20. Fixing screws; 21. Gear lever for 1st and 2nd gear; 22. Cover; 23. Bolt for declutching gear of steering shaft.

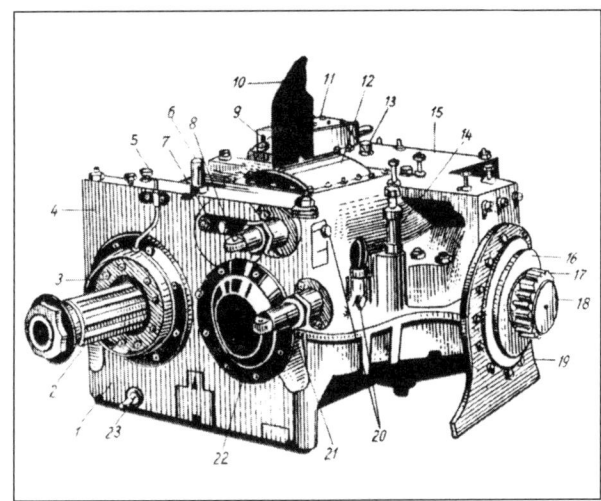

T-34 five-speed gearbox:
Key: 1. Upper housing; 2. Gearshaft housing, 1st and reverse gears; 3. Ventilator socket; 4. Locking screw for oil filler socket; 5. Mount for starter relay; 6. Mount for starter bracket; 7. Grease/oil nipple; 8. Cover of bearing, secondary shaft; 9. Fixator screw; 10. Gearshaft housing, 4th and 5th gears; 11. Lug; 12. Surface for mounting the starter bracket; 13. Spur gear; 14. Seal ring nut; 15. Gearshaft, 2nd and 3rd gear.

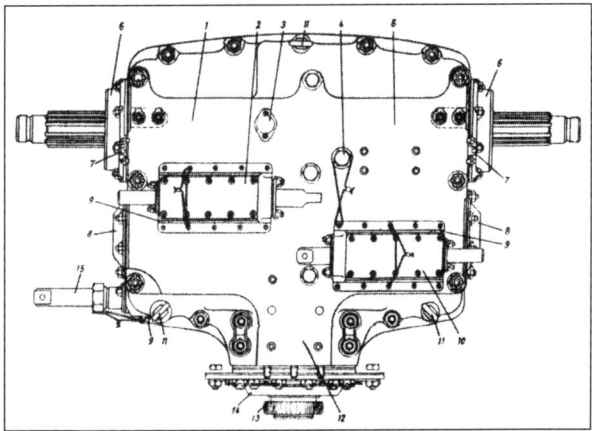

was forced to relocate to Chelyabinsk in the autumn of 1941. The growing demand made the spread of production across four factories necessary from 1941, together with an increase in output.

In 1943 an improved main coupling was introduced and during 1944 a five-speed gearbox. It replaced the simple cogwheel device of four gears which was bad at quick gear changes and was one of the few weaknesses

Production of W-2 Diesel Engines, 1939–45								
Year	1939	1940	1941	1942	1943	1944	1945	*Total*
Number	477	1,933	4,869	16,890	22,955	28,120	20,938	96,182

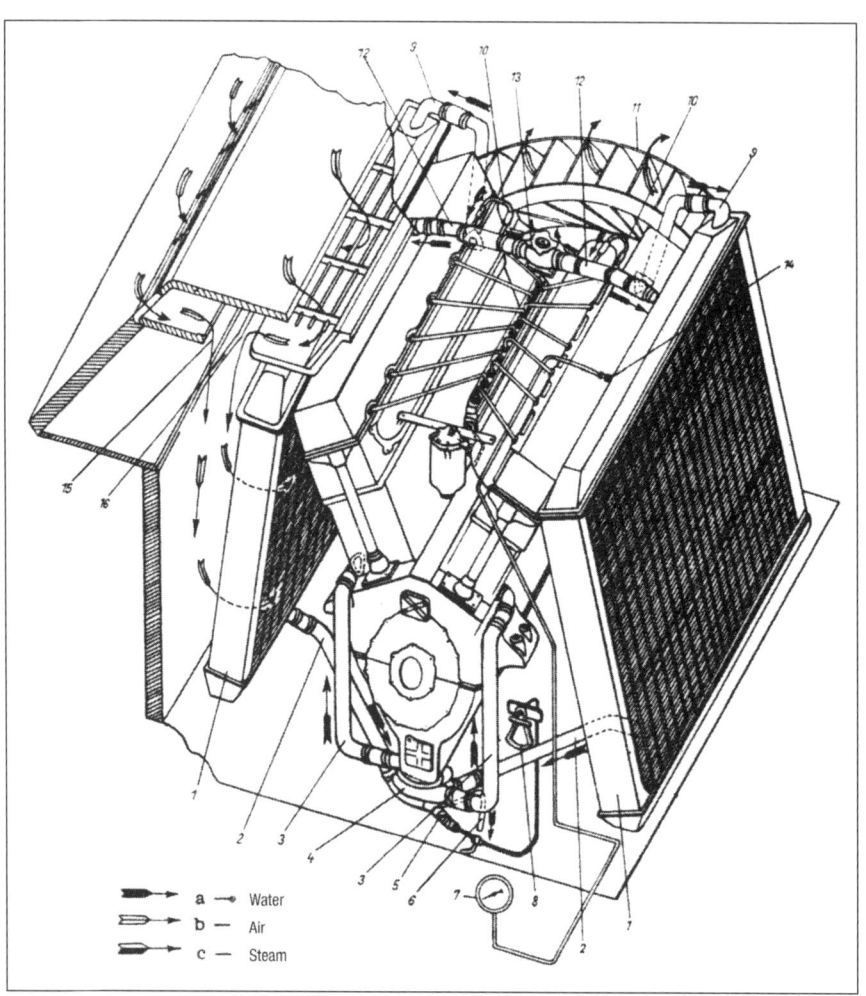

Cooling System of the T-34:

Key: 1. Water cooler; 2. Pipe to pump; 3. Pipes from pump to motor; 4. Water pump; 5. Water release valve; 6. Water release piping; 7. Thermometer for water; 8. Grip for pulling water release valve; 9. Piping for drawing water from motor; 10. Steam pipes; 11. Paddle-wheel ventilator; 12. Pipe for filling water; 13. Water filler connecting piece with steam-air safety valve; 14. Pipe socket for connection to water thermometer; 15. Air intake; 16. Louvres.

---

of the T-34. The new gearbox enabled a better utilisation of specific engine performance, made changing gear easier and reduced the high rate of wear-and-tear on the couplings. Nevertheless the process of changing gear took longer with the five-speed gearbox so that in 1944 new T-34/85s were still being delivered with the old gearbox.

View into the drive compartment of a T-34/85. To the left (between the two exhaust tubes) is the paddle-wheel fan for the W-2 diesel, above it the piping for directing water into the coolers either side. To the right is the five-speed gearbox (partly concealed by the rear armour).

Sights, Observation Equipment and Radios

The sights and observation devices of the T-34 have already been described elsewhere, and in this synopsis only those changes found necessary under battlefield conditions are mentioned. The first series-produced tanks of this type were equipped with the TOD-6 rectilinear telescopic sight with × 2.5 magnification and a 15° visual field. For general observation the gunner/tank commander had at his disposal a PT-6 panoramic periscope with × 2.5 magnification and 26° visual field capable of being rotated through 360° and synchronised with the elevating machinery of the gun. The introduction of the 76.2-mm F-34 L/41.5 gun changed nothing. The old telescopic sight was replaced by the TMFD-7 1941 model, and the PT-6 periscope by the PT-4 or PT-4-14. The optical equipment was of high quality. For the frontal 7.62-mm MG, a PPU-8T telescopic sight (× 1.15 magnification and 26° visual field) was used with fixed crosshairs in the field of vision. The main gun was traversed electrically (the turret took ten seconds to turn). Fine aiming and elevation had to be done manually.

In general the view from the driver's seat with the external vision blocks in a T-34 was poor. In 1942 the turret was fitted with a second PT-4-7

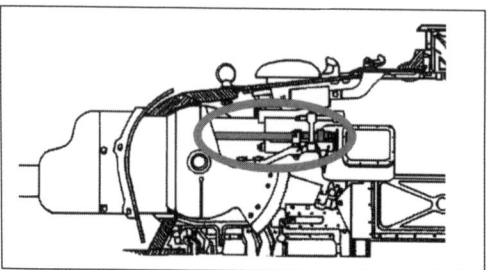
Arrangement of the TOD-6 telescopic sight (× 2.5 and 15°).

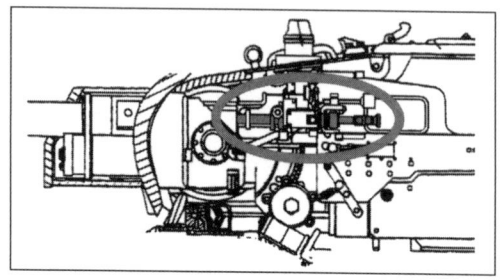
Arrangement of the TMFD-7 telescopic sight (× 2.5 and 15°).

or PT-K periscope on the loader's side. If this was not fitted, the opening for them in the turret roof had to be closed up with a round metal plate. Frequently the supplementary observation equipment was absent from the outset: this had also been the case with the 1941 production in which the panoramic periscope on the left side of the exit hatch in the turret roof had been discontinued. It was not until 1943 when the new cupola for the commander was introduced with periscope and five viewing ports that observation from the T-34 was improved.

1941 model T-34. Hidden by the round metal plate is the aperture for the panoramic periscope in the turret hatch.

Abandoned 1940 model T-34, with the panoramic periscope on the exit hatch of the turret.

T-34, 1942 model, with PT-4-14 panoramic periscope (× 2.5 and 26°) and a PT-K sight on the loader's side.

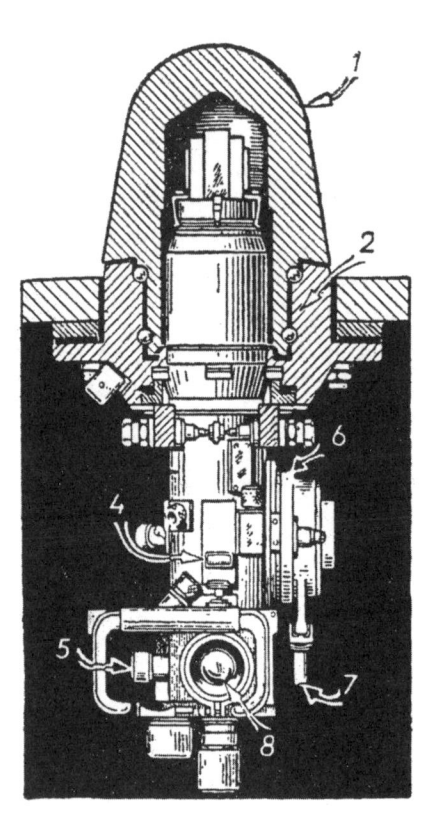 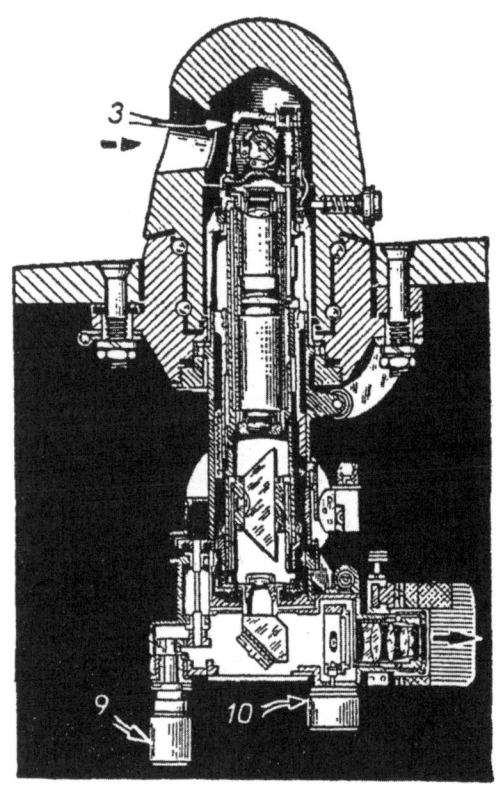

PT-K sight viewed from inside the tank and from the side:
Key: 1. Armoured cap; 2. Operational shell; 3. Head; 4. 360° scale; 5. Knob for side deviations; 6. Setting mechanism for point of aim; 7. Lever for terrain angles; 8. Eyepiece; 9. Setting screw for 360° observation; 10. Screw for sighting angle.

The 71-TK-3 radio transceiver was a standard item of equipment for every T-34. The set could be used for telegraphy or speech and had a maximum range of 7,000 m. It was installed in the hull to the right of the MG gunner. The support for the aerial was outside. Later, the 9, 9P or 9PM sets were used. Command tanks were fitted with more powerful equipment. In the early years of the war there was a shortage of radios as a result of which many T-34s were sent to tank units without means of communication. This made tactical leadership difficult, and was all the worse given the weakness in training and close cooperation between tank crews.

Maintaining Mass Production and Making Improvements

1943 brought the Soviet Union not only a change of course in the war but also a major expansion in the manufacture of war material. Factories located in the eastern regions of the country developed new capacities for production without interruption. The quality of their work improved constantly as did the amounts of equipment of all kinds becoming available. On 23 February 1943, the anniversary of the founding of the Red Army, Stalin declared:

> The Hitler Army, which has the armaments industries of all continental Europe working for it, was until recently superior to the Soviet Union. That was to their advantage. But during the twenty months of the war, the situation has changed. Through the sacrificial devotion of our workers, both male and female, engineers and technicians in the armaments industry of the Soviet Union, in the course of the war the production of tanks, aircraft and guns has grown mightily.

Factory No. 183 Nizhny Tagil delivered 5,684 T-34s in 1942. The tank was of an especially favourable size for mass production. Photo taken in 1942.

Deliveries of T-34/76s 1939–44							
Year	1939	1940	1941	1942	1943	1944	Total
Number	–	117	3,017	12,527	15,833	3,976	35,470

The production of the T-34, to which five factories were committed in 1942 and 1943, had increased fourfold in 1942 as against the previous year. The constantly expanding output made it possible for the Red Army to reinforce its operational tank units. German reconnaissance noted a growing number of independent tank regiments (authorised strength forty-one T-34/76s), tank and mechanised corps (authorised strength 209 and 176 T-34/76s respectively) and tank armies (authorised strength about 850 tanks and SP guns).

The tank losses also grew, as in the Battle of Kursk in the first half of June 1943. Foreign Armies East reported to OKH that in the months of July and August, 4,150 and 4,050 Russian tanks respectively were put out of action on the Eastern Front, and 18,850 for the whole year (65–80 per cent being estimated to be total losses). This high rate of loss left the Soviet

Tanks for the Front. The Soviet armaments industry delivered 28,360 T-34s in 1942 and 1943 alone. In the same period the German armaments industry delivered only 6,913 Panzer IIIs and IVs to the troops .

Assembly line at Factory No. 183 Nizhny Tagil in 1942.

leadership with little room for intervening in tank production, though ideas, designs, projects and prototypes abounded.

Undoubtedly the modernisation of the T-34 and the constant increase of output belongs amongst the major achievements which were brought

A crushed German 5-cm PAK 38 L/60 anti-tank gun. In the front of the T-34 only small areas were vulnerable to hits by the armour-piercing round of this gun, and at less than 100 m. The tank would cover this distance in less than a minute. The rate of fire of the anti-tank gun was 12–14 rounds per minute.

The result of a deliberate collision in which a T-34 model 1942 rammed and pushed aside a German 10.5-cm field howitzer. Its chances of hitting a tank were less with direct fire at short range (about 500 m). Hits with hollow-charge munitions on the other hand had a devastating effect (penetrating 80–100mm armour at 60°).

about far behind the Soviet lines. On the other hand, there were limits to the extent by which the tank's fighting value could be increased. The main concentration of effort was on increasing firepower and improving the armour. The introduction of the five-speed gearbox extended into 1944. The work to improve observation from the tank, linked to the distribution of tasks amongst the four crewmen (the commander also being the gunner) had tactical disadvantages. This was one of the reasons which made it easier for the Germans to engage the T-34 at close quarters especially when some of their other anti-tank weapons had become ineffective.

The new tactic received comprehensive treatment in instruction manuals like H.Dv. 469 *Panzerabwehr aller Waffen* ('Anti-tank Defence by All Weapons'), Issue 4 of 7 October 1942, 'Guidelines for Close Combat against Tanks': 'With the right kind of training and skilful approach, close-combat weapons are suitable for destroying armoured vehicles of all kinds . . . decisive here is the identification of weaknesses . . . which exist in conditions of poor visibility and the limited capabilities of the enemy's close defence.'

Soviet tank crews knew this and paid for it in blood. The removal of the shortcomings was worked on energetically. Responsibility for finding the solution was awarded to Factory No. 183 at Nizhny Tagil, design bureau K8-250 under chief designer A. A. Morozov. In July 1942, the factory came up with a new three-man turret with 52-mm frontal armour. It had a cast and welded commander's cupola with five periscopes, two exit hatches and a PT-4-7 periscope. Together with the now five crew members, ammunition of 77 rounds for the 76.2-mm F-34M L/41.5 gun and 3,600 rounds for the two 7.62-mm DT MGs, the tank with series-type hull weighed 30.9 tonnes. The design was not fully satisfactory, though, because the interior remained cramped. As a possible alternative, in October 1942 Nizhny Tagil produced the T-34S with three-man turret, again with a diameter of 1.420 m, cast-steel commander's cupola and mantlet with cast-steel gun shroud.

Another design, the T-34M, had failed to progress beyond the project phase. It had now been given a new design of hull and armour 60–80 mm thick. The driver's position was transferred to the right and the MG gunner's seat eliminated in favour of a fixed built-in DT MG on the right of the driving seat. The turret had a new exterior with armour 58–80 mm thick but no cupola.

Both above: German 8.8-cm batteries were responsible for a large proportion of the Red Army's tank losses. This photo shows a burnt-out model 1941 T-34 after direct hits, summer 1941.

From October 1942, Morozov's design bureau worked on another project, the T-43 medium tank, of which prototypes were built and tested. The enlarged turret with 90-mm all-round armour weighed 6.5 tonnes. Its characteristics included a high commander's cupola with periscope and four vision blocks plus mantlet with cast-steel gun shroud for the 76.2-mm F-34 L/41.5, 7.62-mm coaxial DT MG and a TMFD-7 rectilinear

T-34 with three-man turret and commander's cupola. This variant weighed 30.9 tonnes and had a top speed of 54.7 km/h.

---

telescopic sight (× 2.5 and 15°). The gunner had the usual periscope sight. The overall height of the tank with new turret was 2.75 m. The hull showed numerous changes to the drive assembly. Armour forward and at the sides of the superstructure was 75 mm, sloping frontally between 36° and 45°, vertical at the sides. The driver's seat, with hatch and two periscopes, was located on the right. The MG arrangement of the T-34M was retained. The operational weight with four crew and a load of 85 shells for the main weapon and 2,772 rounds for the two MGs was 33.5 tonnes.

The reduction in mobility – top speed on the road 48 km/h, cross-country 22.4 km/h – was not reversed by the inclusion of the new five-speed gearbox. The range fell to 280 km even though the fuel carried in the two supplementary tanks at the rear had been increased from 500 to 770 litres. As against the T-34/76, the specific ground pressure rose from 0.72 to 0.91 kg/m². The restricted interior space came in for criticism again and was a reason why the calibre of the main gun could not be increased.

In March 1943 a further development of the T-43 was presented as a variant and tested. An important innovation was the increase of the turret-ring diameter to 1.6 m made possible by enlarging the turret with 90 mm all-round armour. The 76.2-mm gun had a mantlet with cast-steel gun shroud and coaxial DT MG. The turret was fitted with a flat, cylindrical

A T-34 of the 1942/1943 delivery destroyed by shellfire during the Battle of Kursk, July 1943.

cupola with one fixed periscope and another rotatable through 360 degrees, and a second exit hatch with the loader's hatch to the right of it. The ventilator was located at the rear of the turret.

Following the conclusion of testing in August 1943, next month the first 85-mm D-5-T-85 gun became available for installation in the T-34. This gun brought the tank's weight up another 0.6 tonnes to 34.1 tonnes. The 85-mm gun had outstanding ballistic properties. The 9.2-kg shell fired at a muzzle velocity of 792 m/sec would penetrate armour between 78 and 102 mm thick at a range of 1,000 m. Accordingly the T-34 armed with this gun was an important connecting link to the T-34/85. It never entered series production, however, the main reason probably being the new design of hull which would have interfered with the monthly production of medium tanks had it gone onto the assembly lines.

Efforts continued to be made to replace the T-34 with tanks of completely new design. The Kirov Factory at Chelyabinsk – from 1942 the sole manufacturer of the KV heavy tanks – took on a contract to develop a medium tank with thicker armour and requiring less production effort and cost. This resulted in the KV-13 prototype from the drawing board of factory chief designers N. W. Zeiz and N. F. Shashmurin. Cannibalising units and parts from the KV production, the prototype was fitted with the

A model 1943 T-34, out of action in the Dniestr area, July/August 1944. A photo taken at the conclusion of the battle.

In the spring of 1942, the Chelyabinsk Kirov Works developed a model to compete with the T-34. The K-13 had features from the KV heavy tanks but proceeded no further than prototypes and projects.

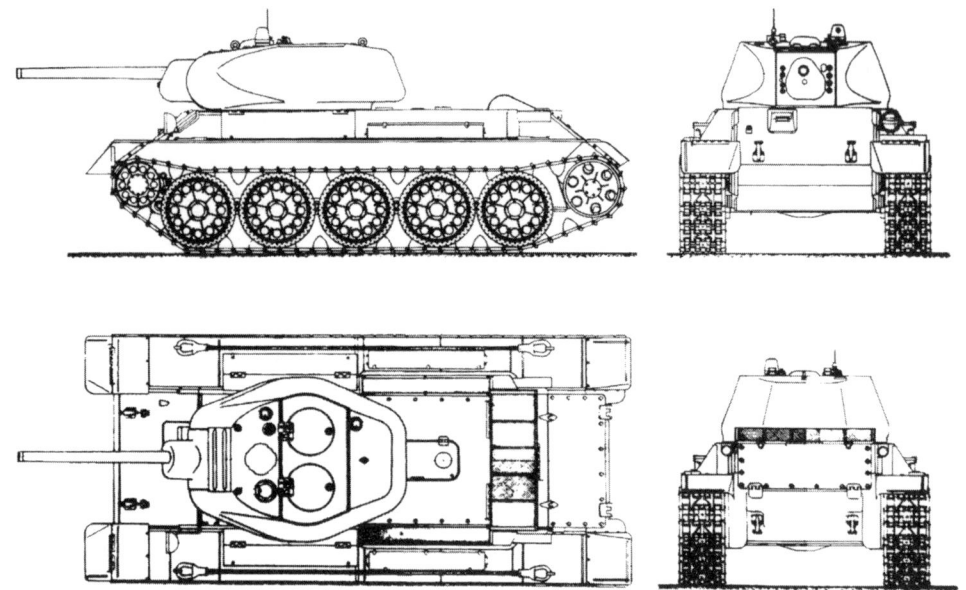

The T-34M project of 1942 in four diagrams.

Cross-section of T-43 with first turret variation, 1942.

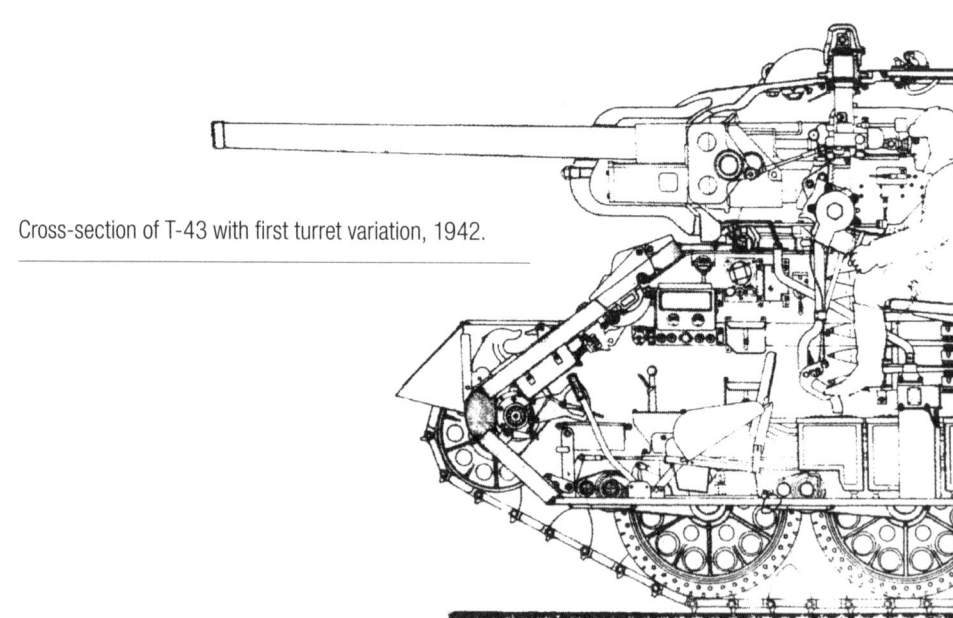

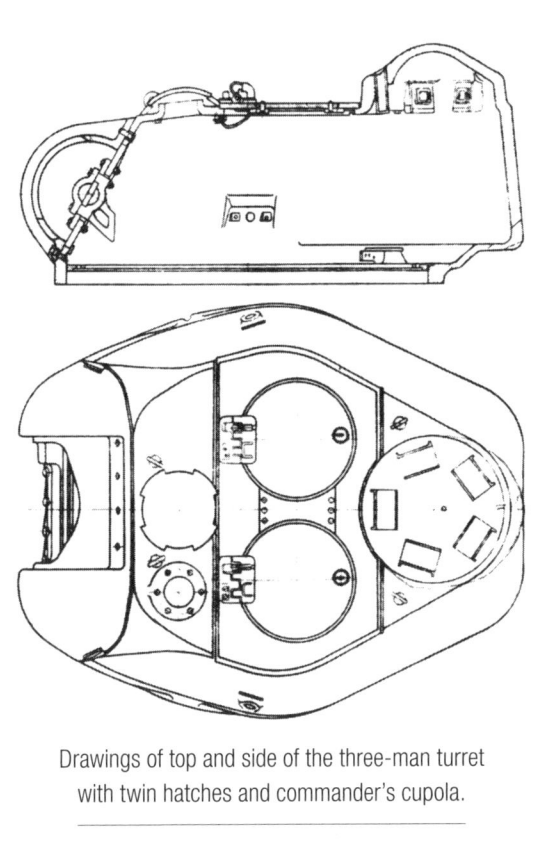

Drawings of top and side of the three-man turret with twin hatches and commander's cupola.

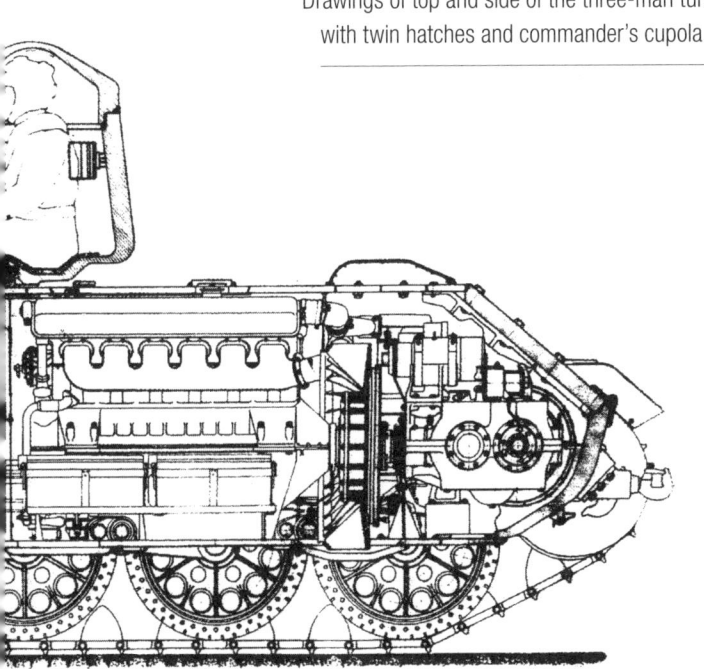

Initial version of the T-43 with 76.2-mm F-34 L/41.5 gun. Testing of the 33.5-tonne tank began at the beginning of 1943. The top speed of 48 km/h was not satisfactory.

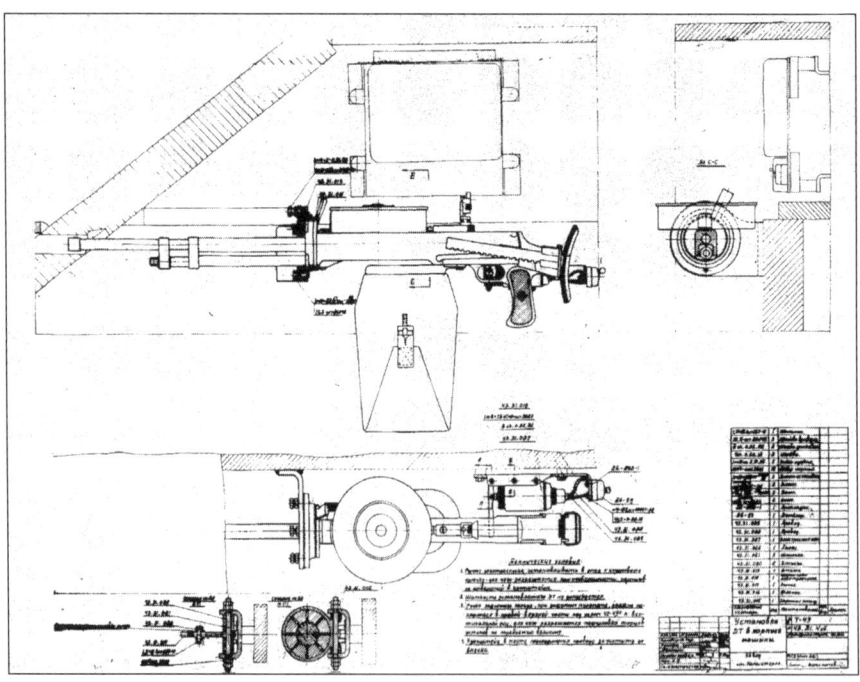

Works sketch of the 7.62-mm DT MG installed in the T-43.

Later version of the T-43 mounting a new turret with a turret ring diameter of 1.6 m.

600-hp 12-cylinder 4-stroke W-2K diesel which with new drive provided the 37-tonne tank with a top speed of at least 65 km/h. The armoured hull, mostly cast steel, had 120-mm armour in the driver's area and between 60 and 85 mm elsewhere. The two-man turret with 85-mm all-round armour housed a 76.2-mm ZIS-5 Model 1941 L/41.5 gun and coaxial DT MG. The three-man crew had a second MG available which could be used as an AA gun if the need arose. Amongst the notable details of the tank were the four pistol ports fitted to the turret for engaging enemy close-combat troops, one on each side and two at the turret rear.

Another variant of the KV-13 which got no further than the project stage had the usual viewing slits in the sides combined with lockable pistol ports and in the rear a second 7.62-mm DT MG. A flat cupola with five periscopes was also to have been incorporated. Another project was a KV-13 with a 122-mm U-11 L/22.7 howitzer installed. There was no series production of the KV-13. Various technical solutions were being worked on for the KV-IS tank completed from August 1942, and so development of T-43 and T-34/85 medium tanks at Chelyabinsk was abandoned and the manufacturing capacity given over entirely to heavy tanks.

Above: In the late summer of 1944, the later version of the T-43 received the 85-mm D-5-T L/51.6 gun. On trials its inferior mobility in comparison to the T-34 was noted.

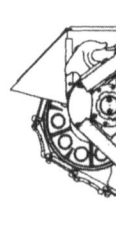

Right: Cross-section of the later version of the T-43 with the new turret. The 76.2-mm F-34 L/41.5 gun was installed as the principal weapon. With a four-man crew, the tank weighed 34.1 tonnes and had a top speed of 48 km/h.

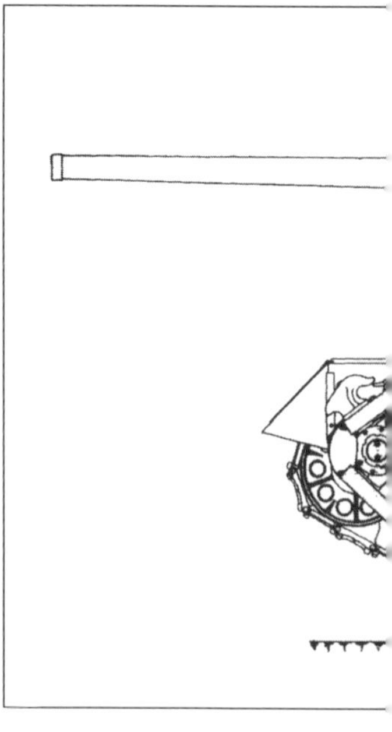

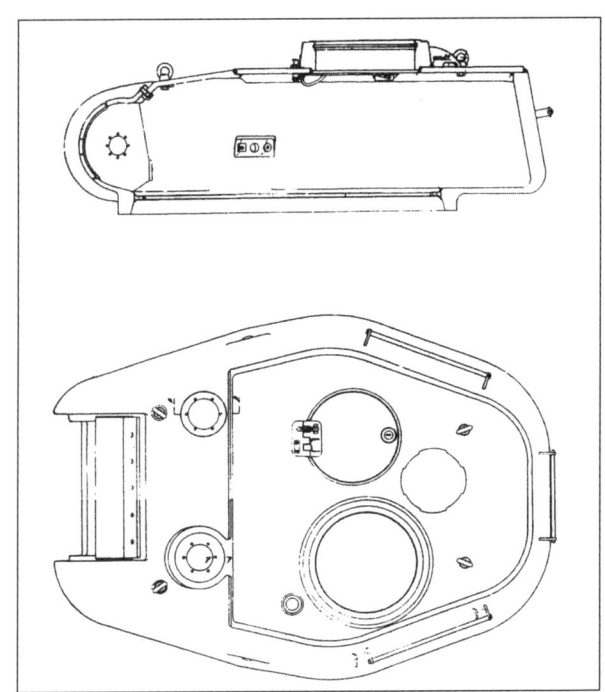

Above: Side and top views of the three-man turret with turret-ring diameter of 1.6 m for the later version of the T-43, spring 1943.

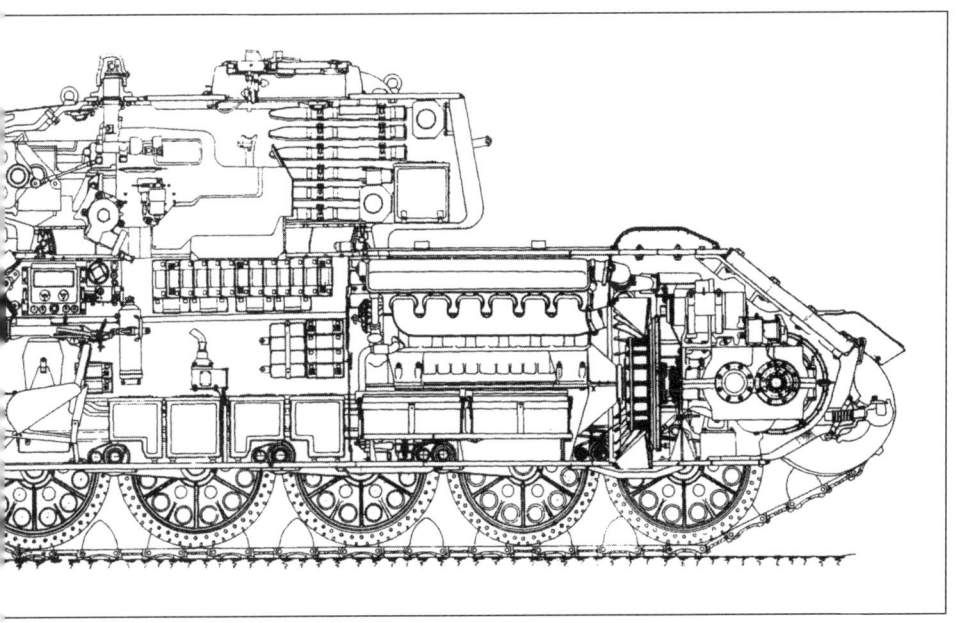

The T-34/85, 1943–1946

In 1943, more powerful panzers and assault guns with 8.8-cm and improved 7.5-cm guns made their appearance on the German side. These were responsible for more than half of the Red Army's losses in tanks that year. Even the T-34 with increased armour had no reliable protection against these weapons from a range of less than 500 m, and could not inflict serious damage on these panzers frontally at ranges over 500 m. All attempts to increase the armour further failed due to the increase in weight and loss of mobility. Therefore the only way of raising the active protection of medium tanks was to increase the firepower, and it now became the aim to develop more efficient guns with improved ammunition so as to engage enemy armour effectively at 1,000 m. Three options were examined:

- In 1943 the 57-mm model ZIS-4M L/73 was installed and tested. The effect of its explosive fragmentation shells against field fortifications was assessed as unsatisfactory, a criticism to which the 76.2-mm guns had already been subjected.
- The use of a longer-barrelled 76.2-mm, the S-54 L/58, was looked at.
- The introduction of an 85-mm gun; LB-1, S-50 and S-53 models were tried as well as the D-5-T.

1943 model T-34/85 with the 85-mm D-5-T L/51.6. Arming the tank with this gun was an interim solution.

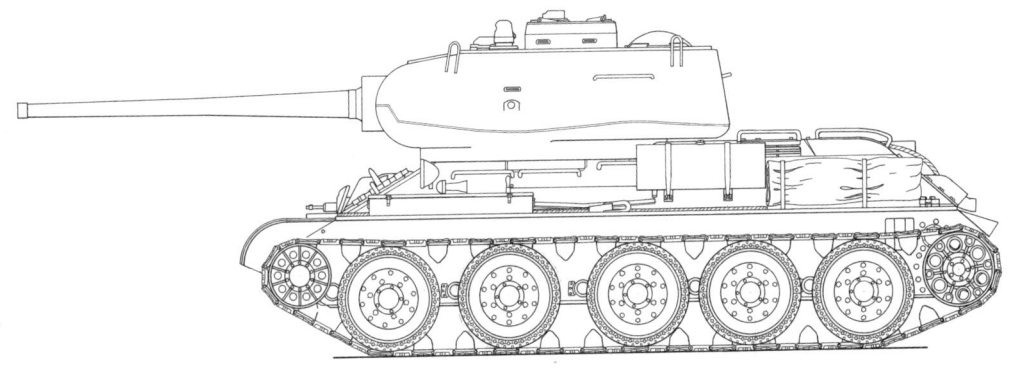

1943 model T-34/85. (*Drawings by Robert Jurga*)

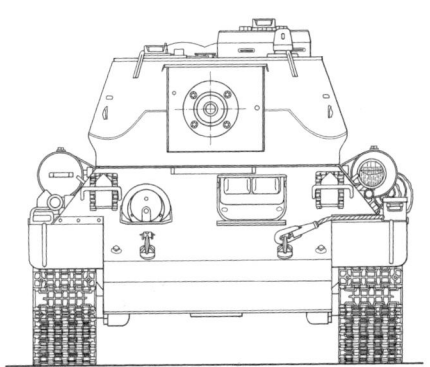

Armour Penetration (mm) at Impact Angle of 60°					
Gun	76.2-mm L-11 L/30.51	76.2-mm F-34 L/41.5			
Range	7.62-cm Pz.Gr. 39 (rot)	BR 350	BR 350 A	BR 350 B	BR 350
100 m	72	–	69–86/80–89	74–89/86–94	92–10?
300 m	–	–	63–79/76–84	65–82/81–90	87–98
500 m	66	–	59–70/70–78	62–76/75–84	77–92
1,000 m	58	58	50–63/63–73	55–71/68–78	–
1,500 m	51	51	43–52/58–65	48–55/62–69	–
2,000 m	–	45	–	–	–

Details from Russian sources are not available: performance is taken from ammunition samples copied in Germany.

Arming the T-34 with an 85-mm gun was first considered in the spring of 1943 after the results of an evaluation of a captured Tiger tank had become available. In April the Defence Committee of the Council of the People's Commissariat supported the suggestion by V. G. Grabin

A photograph from March 1944 showing a 2nd Ukrainian Front tank unit equipped with T-34/85s during the fighting in the Permovaisk area. The tank with the 85-mm D-5-T L/51.6 can be distinguished from the later mass-produced version by the shape of the mantlet.

	85-mm D-5-T L/51.6			
7.62-cm Gr. 39 (*rot*)	BR 365	BR 365 TG	BR 365 K (OG)	BR 365 P
82	97	–	–	–
–	93	–	–	–
75	90	90–105	90–108	100–140
67	80	85–100	78–102	85–118
60	75	78–92	72–90	Shooting forbidden
54	65	72–85	66–82	–

(between 1942 and 1946 Chief of the Central Artillery Design Bureau) of strengthening the T-34 by the use of this calibre.

Factory No. 92 at Gorky, where Grabin had been chief designer until 1942, already had a file of reports from 1940–1 when work had been done on the 85-mm F-30 L/54 gun proposed for the KV-3 (Object 221) and KV-220 (Object 220) heavy tanks. The calibre had proved its worth in the heavy defensive fighting against German panzers in the second half of 1941, but as an AA gun in the form of the 85-mm 52M (M1939) L/54. Therefore various design bureaux were in a position to continue the work begun on this gun in 1938. This earlier experience paid off when the design bureau of F. F. Petrov received the approval of the People's Commissar for Defence, D. F. Ustinov, to begin the development of two variants of an 85-mm gun to be installed in 1943–4 in the KV-85 and IS-1 heavy tanks, and also the SU-85 assault gun. A feature of the design was having the barrel recoil brake and pneumatic recuperator above the barrel, which limited the degree to which the gun could be elevated, and therefore required modification for the T-34. The design bureau of A. Savin proposed a third variant at Factory No. 92, and in the late summer of 1943 prototypes were tried out at the Gorochovetz firing range, first with a T-34/76 of the current year fitted with a two-man, 1.42-m diameter turret. It was found that the lack of space in the gun compartment hindered the crew, especially when loading the 12.6-kg shells which were between 89 and 92 cm long (previously 9.2 kg

1943 model T-34/85 destroyed by shellfire. Photo from spring 1944.

and 68.6 cm long). The required rate of fire of eight to ten rounds per minute could not be achieved. Better results were obtained on the later version of the T-43 with a turret turntable diameter of 1.60 m, but in any

Armour Penetration of the 85-Mm D-5-T-85 and ZIS-S-53-85

Shell	BR 365		BR 365 K		BR 365 P (sub-calibre)	
Muzzle velocity	792 m/sec		792 m/sec		1,040 m/sec	
Angle of impact	60°	90°	60°	90°	60°	90°
Range						
100 m	96	117	102	124	100–122	140–155
300 m	92	113	95	116	112	150
500 m	89–90	109–110	88–90	108–110	100–102	137–140
1,000 m	80–81	99–100	75	90–92	79–80	107
1,500 m	74	90	63	76	61	81
2,000 m	65–67	80–82	50–52	64–65	–	–
3,000 m	55	67	37	45	–	–

1944 model T-34/85 delivered from Factory No. 183 Nizhny Tagil.

case it was clear that the T-43 would not enter series production for it would have caused setbacks in the production of medium tanks.

A compromise had to be found to modify the two-man turret sufficiently to fit an only slightly adapted T-34 hull. This task was taken on by Factory No. 112 Krasnoye Sormovo at Gorky where, within a short time, a turret was devised with a 1.57-m diameter turret ring with 90-mm frontal, 75-mm lateral and 60-mm rear armour as on the T-43. Series production was prepared, over-hastily as it proved. Testing of various models of the 85-mm gun revealed a variety of shortcomings such as insufficient elevation (-5°/+22°) and a slow rate of fire. Moreover, it was not possible to load and fire the gun when the tank was on the move, although the process was fairly rapid when stationary.

The S-53 from the Factory No. 92 design bureau was especially criticised, and heated exchanges occurred between the two manufacturers as to whether the turret turntable or the gun had to be modified. The decision was referred to higher authority, and as a result the S-53 underwent changes and the turret diameter was increased to 1.6 m. The S-50 model was rejected completely because it required non-standard ammunition.

In December 1943 the first two T-34s were delivered to Factory No. 92 for fitting with the new turret. On 15 December the Defence Committee decided to introduce the T-34/85 into service. Since problems with the 85-mm S-53 gun persisted – only twenty-one had been completed by the end of the year – the 85-mm D-5-T L/51.6 was used as a provisional solution. This gun weighed almost 400 kg more, had restricted elevation, and could only fire five to eight rounds per minute to a range of 12,900 m, but almost 300 of them were ready for immediate installation. Between January and April 1944, Factory No. 112 at Gorky delivered 255 T34/85s and five command tanks of the same type fitted with the D-5-T, the TSch-15 telescopic sight (× 4 and 16°) and the PT-4-15 periscope (× 2.5 and 26°).

The development of the desired 85-mm gun was undertaken by the Central Artillery Design Bureau in October 1943, and on 1 January 1944 the 85-mm ZIS-S-53-85 was designated as the main armament of the T-34/85, mass production being authorised in March 1944. In the following month it replaced the D-5-T. The 85-mm ZIS-S-53-85 was outstanding for its simple construction, low manufacturing cost and high reliability. Its rate of fire and range exceeded that of the D-5-T. In the year of its introduction, its performance met the requirements for battle against German panzers and self-propelled guns, the sub-calibre shell (BR-365P) being effective

D-5-T-85 L/51 and ZIS-S-53-85 Guns Manufactured

Year	Factory	85 mm D-5-T-85 L/51.6	85 mm S-50, S-53, ZIS-S-53-85 L/54.6	Totals
1943	No. 9	283	–	304
	No. 92	–	21	
1944	No. 9	160	–	11,778
	No. 8	100	–	
	No. 92	–	8,180	
	No. 13	–	3,338	
1945	No. 92	–	10,060	14, 265
	No. 13	–	4,205	
Totals		543	25,804	26,347

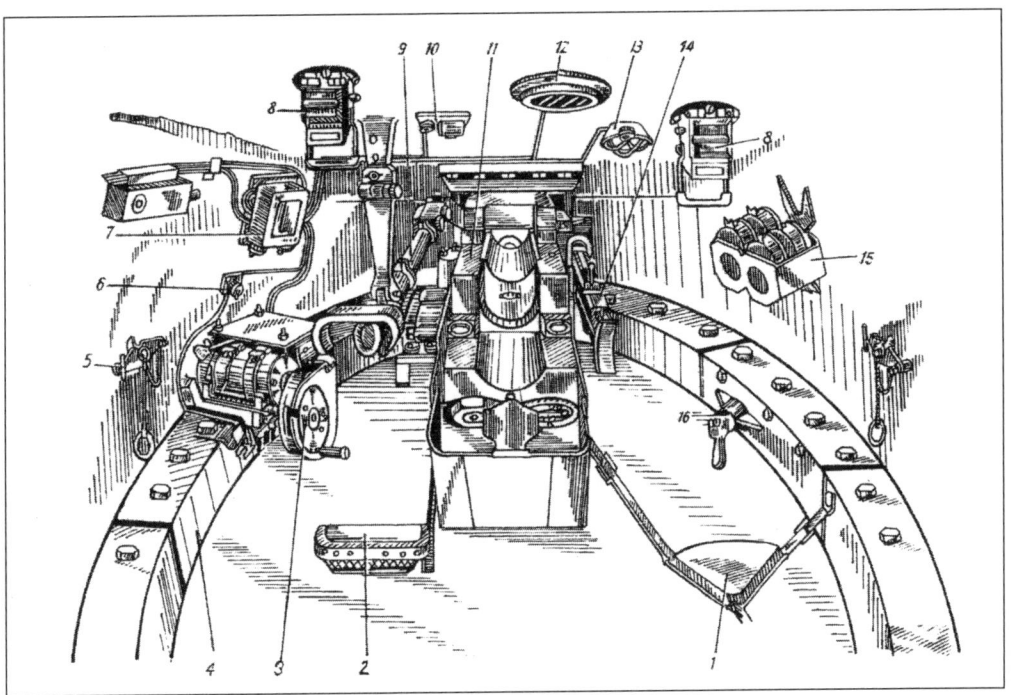

Fighting compartment of the T-34/85:
Key: 1. Loader's seat; 2. Gunner's seat; 3. Traversing machinery; 4. Turret turntable panelling; 5. Lock for pistol port; 6. Lighting switch; 7. Switchboard for turret electricals; 8. Periscope; 9. TSch-16 telescopic sight; 10. Switchboard for TSch-16 telescopic sight; 11. 85-mm gun; 12. Ventilator; 13. Turret lighting; 14. DT MG; 15. Rack for MG drums; 16. Turret lock.

at ranges up to 1,000 m. The design bureau at Factory No. 183, led by A. Moloshtanov and M. A. Nabutovski, was responsible for the final form of the turret.

The gun, 7.62-mm coaxial DT MG and TSch-16 telescopic sight (× 4 and 16°) were integral to the mantle of the turret, the gun with minor variations of shroud. The PT-4-15 periscope (× 2.5 and 26°) on the turret roof in front of the commander's cupola was discarded since observation for the tank commander had been much increased. The cupola had four lockable viewing slits with a wide visual field and a rotatable Mk 4 periscope. Two other Mk 4s were situated in the turret roof, one to the right of the loader, another on the left side for the gunner. The commander's exit hatch was in two parts on those tanks manufactured at Factories 112, 174 and 183 on the earlier completions runs: in the later runs it was a single part.

The installation of two ventilators in the turret roof, as a rule at the rear of the turret slightly to the right, was an improvement made for the benefit of the crew. In 1944–5, Factory No. 112 delivered T-34/85s having turrets equipped with one ventilator forward and one at the rear. Various configurations are known aboard T-34/85s. These were always a compromise during the effort to make the armament more efficient. The increase in weight was kept within bounds, the height of the tank being increased by 30 cm from 2.4 m (model 1943 T-34/76) to 2.7 m. The crew had more room and a better all-round view. In order to enlarge the interior, the designers were forced to increase the space all around the turntable; this space was especially vulnerable if the tank were hit. The large extension at the rear was necessary to accommodate the supports for twenty 85-mm ready-use shells (four of them on the right-hand inner turret wall.)

The five-speed gearbox was not yet fitted in all T-34/85s; the transition took time. Various other changes included supports for the additional fuel tanks. The need to improve the armour protection of the T-34/85 hull was passed to Factory No. 183 Nizhny Tagil which, after examining several variants, decided to lengthen the hull and alter its form, and thus boost the frontal armour to 75 mm. These changes increased the weight of the tank to 32.23 tonnes.

Of the 13,618 medium tanks delivered in 1944, 10,632 were T-34/85s, around 2,000 fewer overall than the previous year. Instead the fighting front received more than 2,400 self-propelled guns based on the T-34.

Very soon the responsible authorities wanted a more powerful gun for the T-34/85. On 10 December 1943, the 85-mm S-31 BM and S-50 (muzzle velocity 920 m/sec) were tried out on the firing range. They were rejected for their complicated and expensive design and the numerous defects which came to light during the testing. The basis for all future work was now a decision of the Defence Committee of 27 December 1943 in which the idea of an improved gun system for the medium tank was raised once again, including an 85-mm gun with a muzzle velocity of 1,000 m/sec.

In cooperation with the Central Artillery Design Bureau, the D-5-T 85-BM (muzzle velocity 920–950 m/sec) was offered for the IS-85 heavy tank and in September 1944 the ZIS-85-PM (muzzle velocity 950 m/sec) for the T-34/85 medium tank. These guns did not pass muster on the firing range, and other projects were also rejected such as the 85-mm WP (muzzle

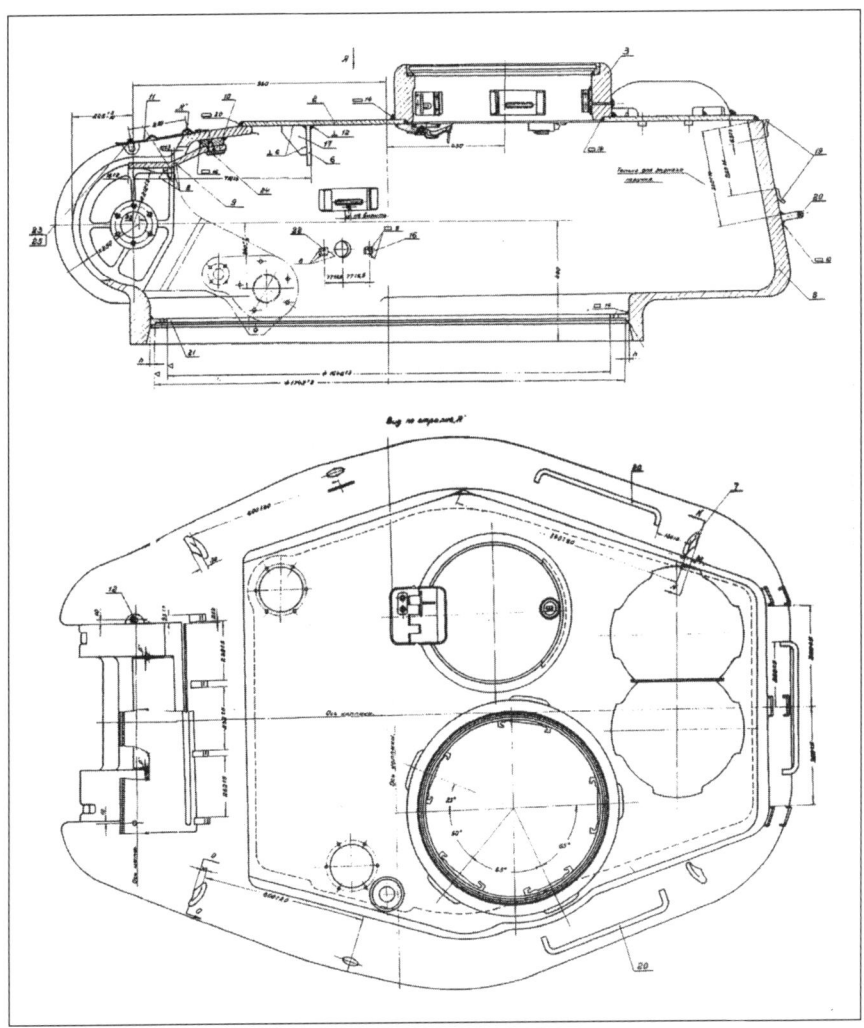

Works drawing of the three-man turret for the T-34/85 with 85-mm ZIS-S-53 L/54.6 gun.

velocity 920 m/sec) and W-9-k (muzzle velocity 1,150 m/sec), the latter with a tapering barrel. At the beginning of 1945 an 85-mm ZIS S-53-T-85 was tested with an extension added to the barrel for a length of 5.42 m (L/64). This weapon received the designation ZIS-1 and fired heavy sub-calibre shells weighing 4.59 kg at a muzzle velocity of 1,093 m/sec, and AP shells at a muzzle velocity 800–900 m/sec. This development got no further: the main point of interest was a larger calibre as the main armament for medium tanks.

The weight of the 85-mm ZIS-S-53 L/54.6 is stated as 1,050 kg, about 350 kg lighter than the D-5-T L/51.6.

Consideration had previously been given to arming heavy tanks with a 100-mm gun. What was lacking was a suitable barrel and ammunition. In order to save time, in the autumn of 1943 the designers returned to the 100-mm B-34 L/58 naval gun. The calibre offered a number of advantages if installed in a medium tank – a further increase in penetration of armour and a greater explosive and splinter effect against fortifications. In July 1944 a contract was issued to install a 100-mm gun in a T-34/85. From the design point of view this presented a challenge. The existing diameter of turret ring was not sufficient, the requirement – as for the IS heavy tank – being 1.8 m at least, which meant a completely new hull. A way out was seen in the 100-mm ZIS 100, for which the 85-mm ZIS-S-53-T-85 was the basis. The test in a T-34/85 from the series production proved a failure.

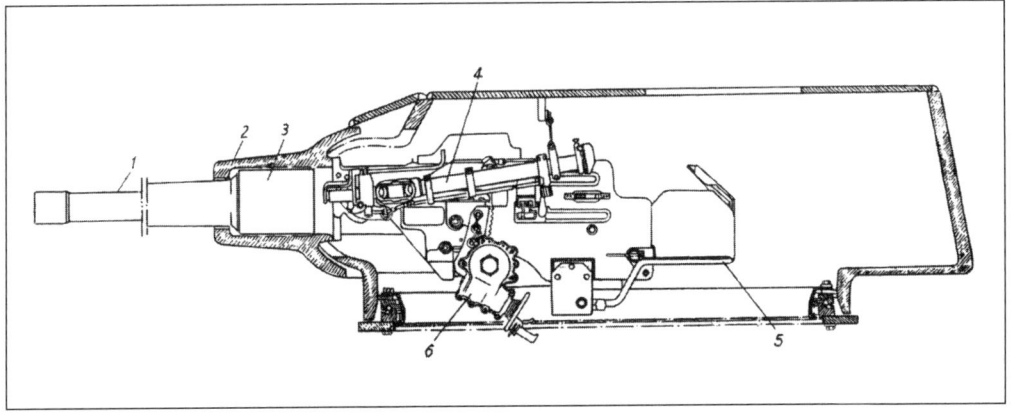

Above: 85-mm ZIS-S-53 L/54.6 gun viewed from the left:
Key: 1. Barrel; 2. Mantlet; 3. Barrel cradle; 4. TSch-16 telescopic sight; 5. Cartridge deflector; 6. Elevation machinery.

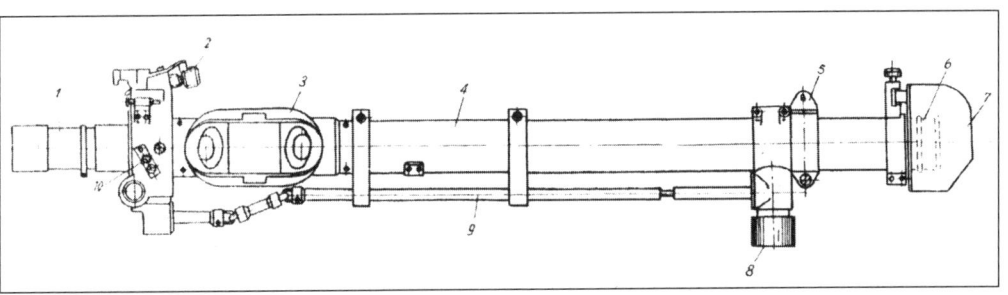

Above: TSch-16 telescopic sight (× 4 and 16°):
Key: 1. Head piece; 2. Securing screw; 3. Hinge; 4. Main tube; 5. Supporting collar; 6. Viewing socket; 7. Forehead rest; 8. Tapping screw; 9. Gimbal shaft; 10. Adjusting lever.

Right: Comparison of sizes of Soviet tank ammunition, left to right: 7.62-mm; 14.5-mm; 20-mm; 45-mm; 76.2-mm; 85-mm; 122-mm shell and cartridge casing.

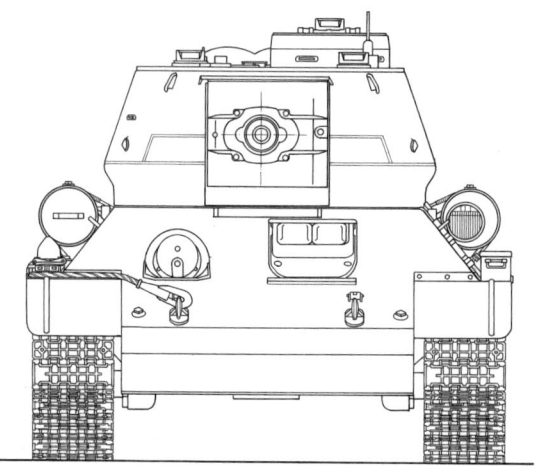
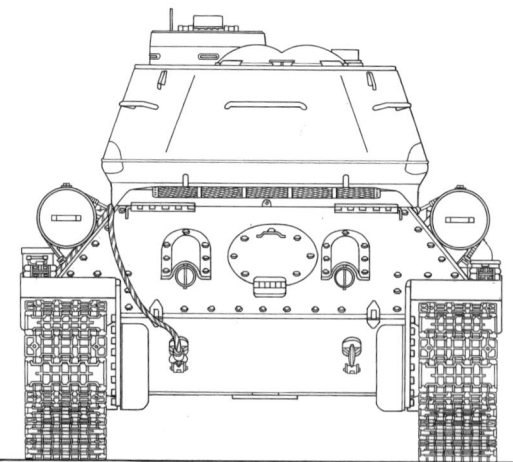

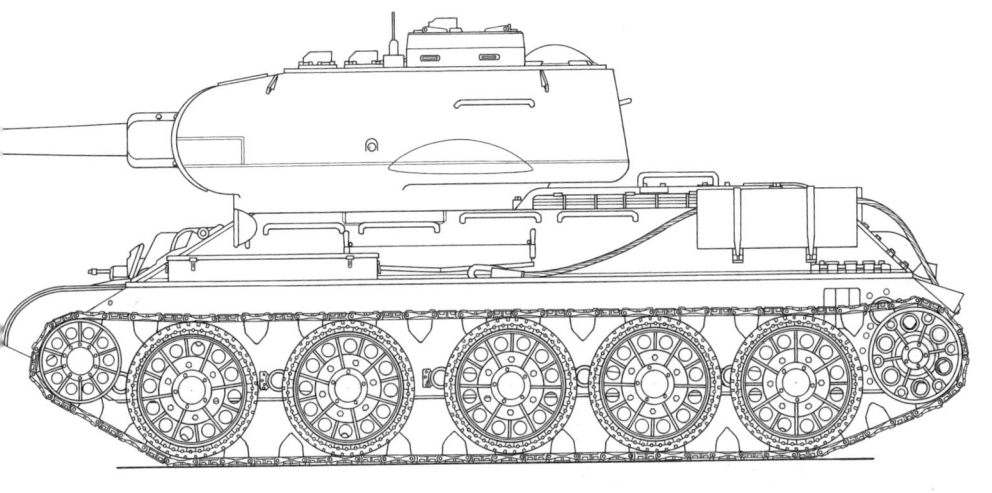

1944 model T-34/85. (*Drawings by Robert Jurga*)

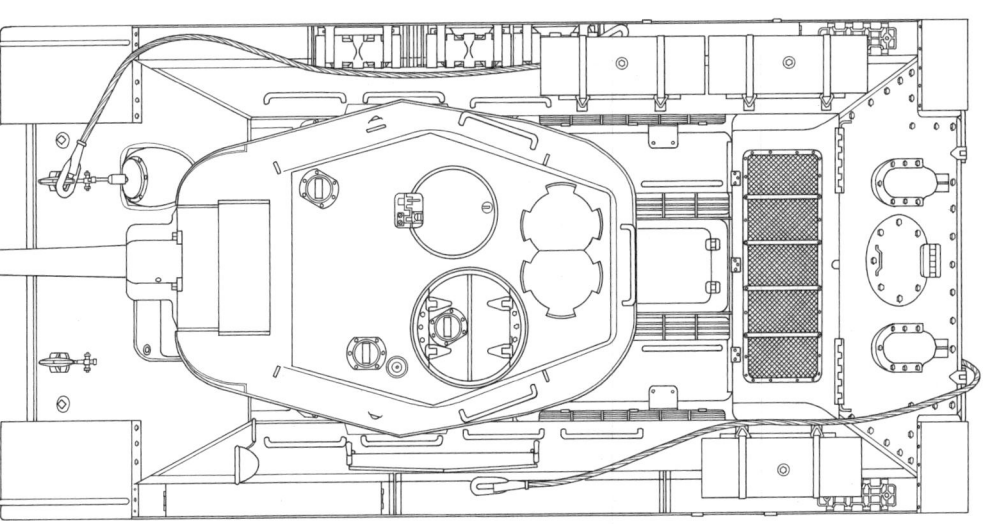

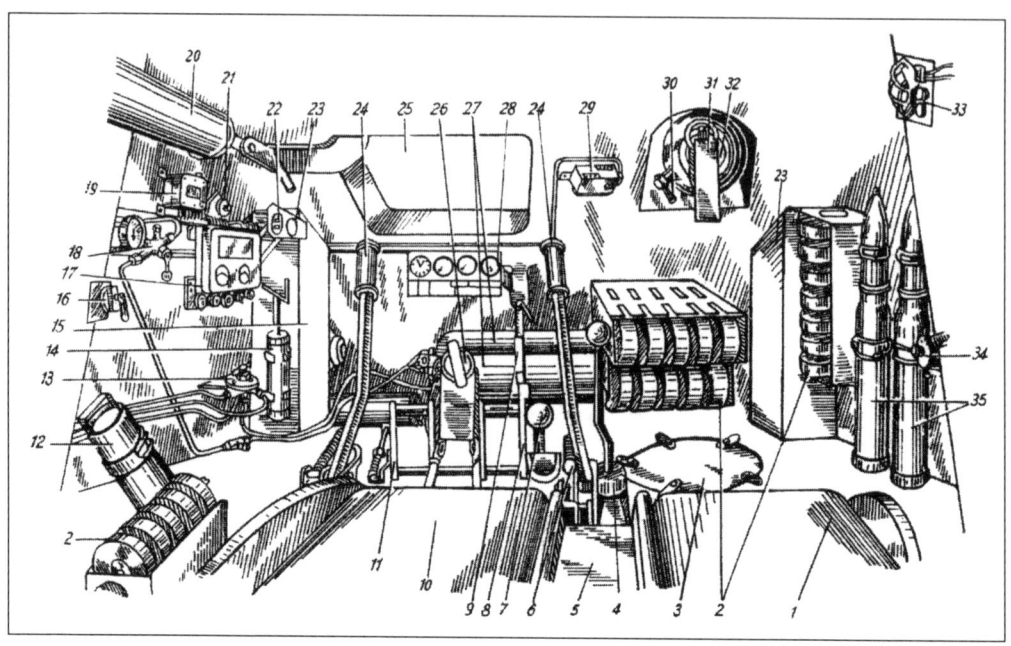

T-34/85 driver's compartment:

Key: 1. MG gunner's seat; 2. MG ammunition drum racks; 3. Emergency exit hatch; 4. Gear change lever; 5. Tool box; 6. Hand control knob; 7. Accelerator pedal; 8. Brake pedal; 9. Stop pawl to brake pedal; 10. Driver's seat; 11. Clutch pedal; 12. Fire extinguisher; 13. Air distribution valve; 14. Grease gun; 15. Suspension shaft for forward road wheel; 16. Ventilation valve for fuel supply system; 17. Switchboard for electrical instruments (reduction valve for compressed-air starter system); 19. Control switch RRA-24F; 20. Adjusting mechanism for driver's hatch cover; 21. Starter button.; 22. Revolutions gauge; 23. Speedometer; 24. Steering levers; 25. Driver's hatch; 26. Hand air pump; 27. Compressed air bottle; 28. Instrument panel; 29. Intercom No. 3 for driver; 30. Lashing for bullet shield; 31. Hull MG; 32. Shield for hull MG; 33. Switchboard for emergency lighting; 34. Ground switch; 35. Ready ammunition in holding clips.

Even the use of a single-chamber muzzle brake could change nothing: the recoil was too powerful.

After the failure of the ZIS-100 trials in February/March 1945, the designers at Factory No. 92 took a fresh step forward with the 100-mm LB-1. They succeeded in reducing the recoil to such an extent that the chassis vibrations were reduced to an acceptable level. The telescopic sight fitted was the TSch-19. A series-built T-34/85 weighed 30.8 tonnes with the gun and had room for 30 shells and 1,953 rounds for the secondary armament. However, the elevation was only −2° to + 20° and the rate of

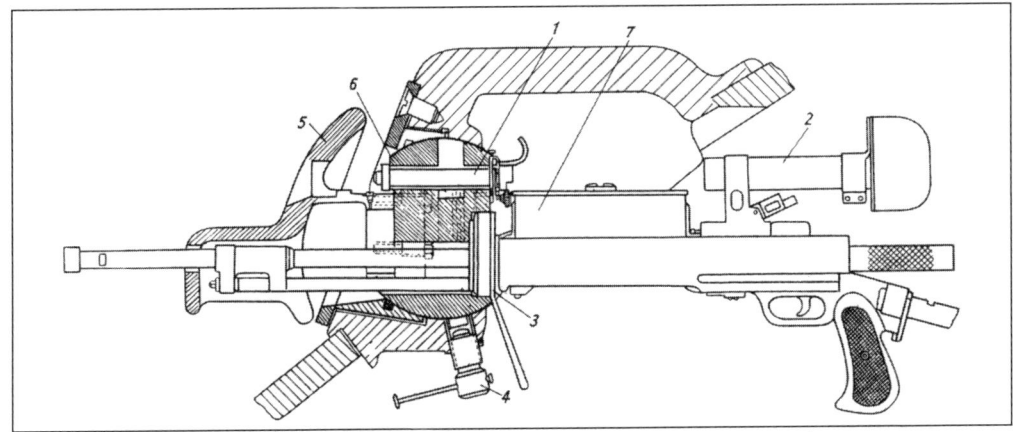

Hull 7.62-mm MG:
Key: 1. Forward tube of PPu-8T telescopic sight; 2. Rear of tube; 3. Clamping ring; 4. Lashing of shield; 5. Protective armour plate; 6. Shield; 7. Drum magazine.

fire was insufficient. The 100-mm D-10-T L/53.5 from design bureau KB-250 at Nizhny Tagil was now offered as a competitor. In order to fit this gun in the T-34, a new flat turret with a turntable diameter of 1.7 m was developed. It projected beyond the armoured structure of the hull forward and on both sides. The elevation of this gun is reported to have been −3° to +18°, the rate of fire four to six rounds per minute. Using the 15.4 kg BR 412 heavy AP shell, 110 mm of armour could be penetrated at

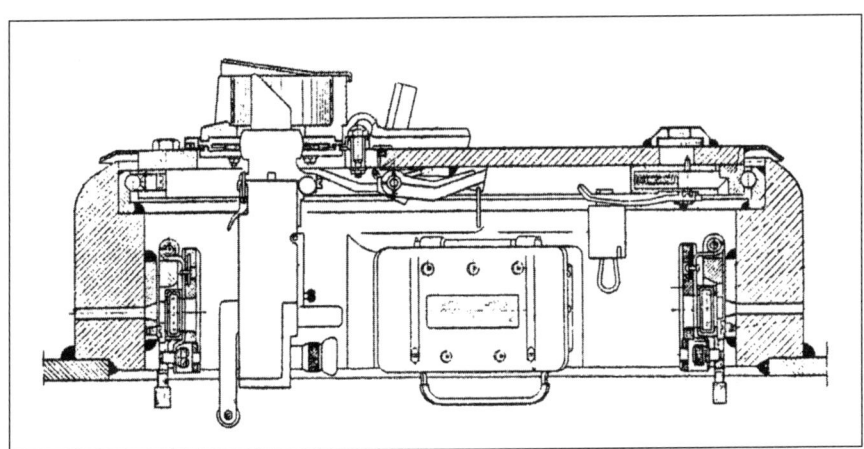

Cross-section of commander's cupola of a T-34/85.

Finished tank, Factory No. 174 Omsk.

A T-34/85 with damaged mudguards. This was a common occurrence.

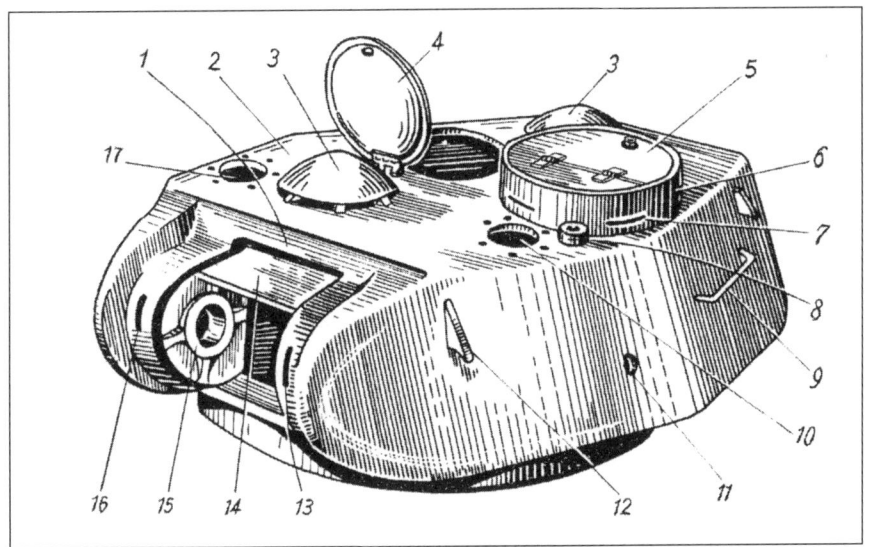

Turret from a tank built at Factory No. 112 Gorky. Identifying feature: the arrangement of armoured covers for the two ventilators

Key: 1. Turret housing; 2. Turret roof; 3. Armoured domes over ventilation apertures; 4. Hatch cover; 5. Commander's cupola cover; 6. Commander's cupola; 7. Viewing slits; 8. Base for aerial; 9. Handgrip for tank riders; 10. Aperture for gunner's periscope; 11. Pistol port; 12. Hook for lifting off turret; 13. Slit for the TSch-16 telescopic sight; 14. Shielding; 15. Spigot support; 16. Slit for MG barrel; 17. Aperture for loader's periscope.

an impact angle of 60° from a range of 1,000 m, but the drawbacks outweighed the advantages.

Substantial changes were made to the hull design: with a four-man crew, 30 rounds for the main gun and 1,900 for the secondary armament, the weight was 33 tonnes and the top speed dropped to 48 km/h. Medium tanks of the T-34/100 type did not enter series production because the T-44 was available as a successor and had greater potential, but the 100-mm D-10-T-L/53.5 was now settled on as the future primary weapon.

The development of the new medium tank, the T-44, was begun in Factory No. 183 in the course of 1943 under the direction of Chief Designer A. A. Morozov. Elements of the T-34 and T-43 flowed into the design. In November of that year, workers and technicians at the Nizhny Tagil experimental workshop assembled the first T-44 hull. The Red Army had retaken Kharkov five months previously and, according to the plan, series production of the T-44 was to have been started in January 1944 at Factory

The numerous captured T-34/76s and T-34/85s played an important role for the Germans, and were designated Pz.Kpfw. 74/76(r) and Pz.Kpfw. T34/85(r) respectively. An example of the latter is seen in this photo taken in 1944 as a member of the 'Jaguar' special unit.

A T-34/85 in the outskirts of Berlin, end of April 1945. These tanks were ubiquitous. On 2 April 1945, Fourth Guards Tank Army had over 387 tanks and SP guns of which 255 were T-34s.

A T-34/85 out of action after receiving hits on the exhaust and motor compartment covers, mantlet and gun shield.

Both above: Wrecked T-34/85s at Harta/Saxony during the fighting in the spring of 1945.

T-34/85s Completed 1944–1 July 1945			
Year	Factory	Numbers	Totals
1944	No. 183	6,583	
	No. 112	3,079	10,632
	No. 174	970	
1945	No. 183	3,759	
	No. 112	1,490	5,959
	No. 174	710	
Totals			16,591*
* Excluding OT-34/85 flamethrower tanks.			

No. 75. Up to that time, however, it had only been possible to test the first prototypes with T-34/85 turrets but even so it was seen that the tank had a whole series of new design features:

- Torsion bar suspension.
- 500-hp 12-cylinder 4-stroke W-2/IS diesel installed crosswise with five-speed gearbox. The W-2/IS was replaced later by the equally powerful W-44.
- A flatter hull, more spacious in comparison with the T-34. The driver's position was on the left side, recognisable by the rather higher superstructure with large viewing port at the front, exit hatch in the roof (with two fixed and one rotatable Mk 4 periscopes). To the right of the driver were the 7.62-mm DT MG, then the intercom system, then racks for the gun ammunition. In the further course of development the hull underwent a number of changes, also so as to accommodate other designs of turret with a more powerful armament.
- Initially the T-34/85 turret was installed. At short intervals the manufacturers provided more efficient tank guns for testing in suitable turrets, the frontal armour of which was between 115 and 120 mm.
- In order to achieve one of the main aims, increasing the firepower of medium tanks, various guns were tried; the 85-mm D-5-T-85 L/51.6 and ZIS-S-53 L/54.6, the 100-mm LW-1 and

Soviet and Polish tank formations suffered heavy losses on German soil in the course of fighting in the spring of 1945. Third Guards Tank Army had 198 T-34s damaged and lost 97 more (more than 50 per cent of them victims of Panzerfausts) between 15 April and 2 May. Second Polish Army lost 160 tanks (57 per cent of its complement as of 16 April 1945).

D-10-K L/53.5, and also the 122-mm D-25-44. For the series production of the T-44, the ZIS-S-53 was chosen.

The weight of the first T-44 reached 30.4 tonnes, of the T-44/100 33.5 tonnes (which was not wholly attributable to the heavy armament and turret armour). The hull armour forward was increased from 75 mm to 90 mm. An improvement in mobility was not achieved for the T-44A (Object 86) until the fitting of the 500 hp W-44 diesel when the tank achieved a top speed of 60.5 km/h. The range remained unsatisfactory at 190 km. It is noteworthy that the overall height of all T-44 modifications was 27 cm below that of the T-34 although at 3.1 m it was slightly broader. In 1944, Factory No. 75 delivered 25, in 1945 240 and until the termination of production in 1946 another 390. As its successor, the T-54 (Object 137) was ready to lead the post-war era of Soviet tank building.

A T-44 with the 85-mm ZIS-S-53 L/54.6 gun. The 31.8-tonne tank had a four-man crew. Factories No. 183 Nizhny Tagil and No. 264 Stalingrad (Southern Boat Yard) were involved in its manufacture. The 520-hp diesel provided the vehicle with a top speed of 51 km/h.

The T-34/100 version from Factory No. 183 was several tonnes heavier than earlier variants. The 100-mm LB-1 gun was fitted in a newly designed turret with a turntable diameter of 1.7 m. With a four-man crew the tank weighed 33 tonnes, but could only manage 45 km/h.

A T-34/100 with the 100-mm LB-1 gun from Factory No. 92. With a four-man crew it weighed 30.8 tonnes and had a top speed of 55 km/h.

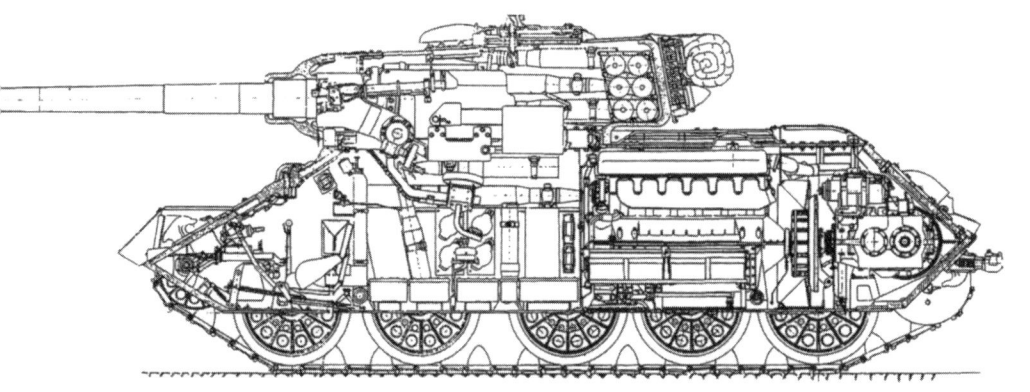

Cross-section of a T-34-100 (Factory No. 183 Nizhny Tagil).

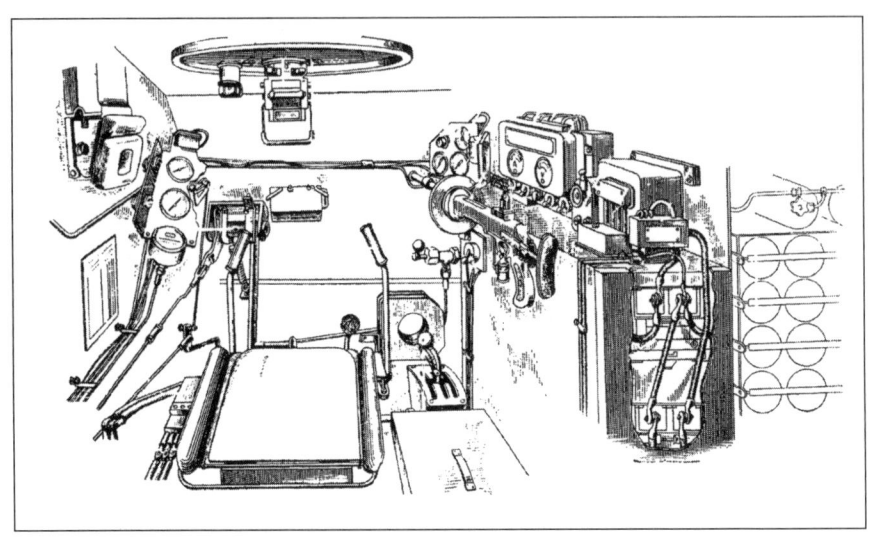

The T-44 driver's compartment. On the left the seat for the mechanic/driver, on the right the 7.62-mm DT MG.

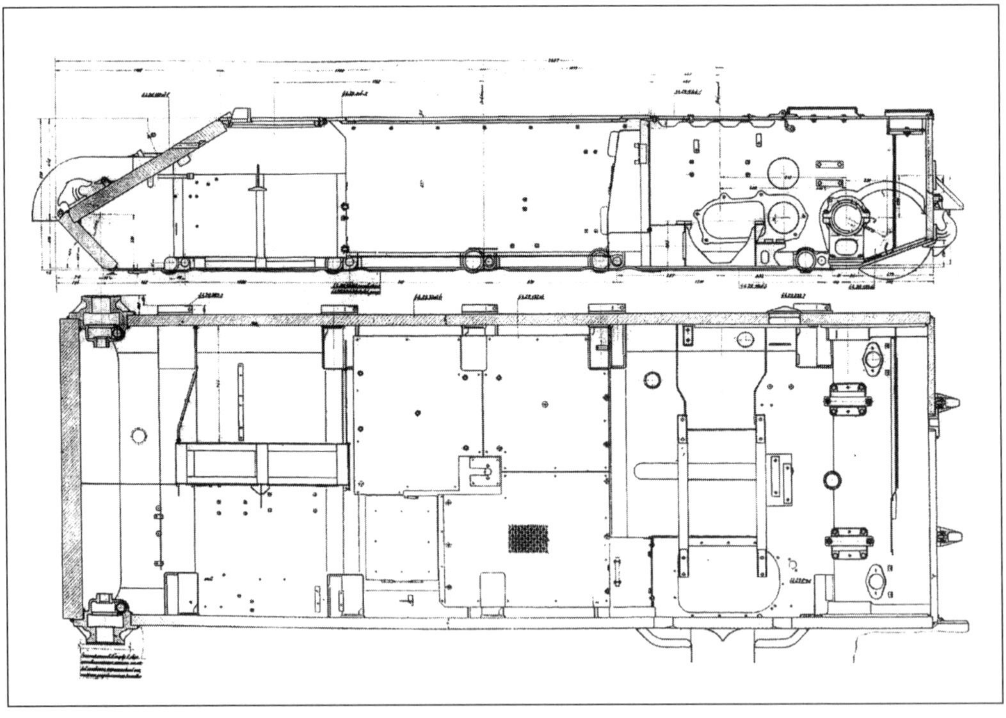

The hull for the T-44, as of July 1944.

Removal of the 12-cylinder 4-stroke W-44 diesel engine.

Removal of the five-speed gearbox.

The T-44 tank, as of the end of 1944. The driver's hatch is fitted with a periscope (between gun mantlet and headlamp).

3 Specialised Variants

As already mentioned in Chapter 1, before the outbreak of the Second World War, the Soviet Union had by far the largest fleet of tanks and armoured vehicles of all kinds in the world (as of 22 June 1941, 30,120 not counting the losses in Spain, Manchuria and Finland). The Great Patriotic War changed nothing. By 1945, 112,500 tanks and self-propelled guns had been delivered, among them 59,000 T-34s. Compared with other tank-building nations, the variety of specialised tanks was not large, though their numbers were.

Before the war, the Red Army experimented with special tanks on a large scale. These included command, flamethrower, remote-controlled flame-thrower/explosive, mine-clearance, submersible, amphibious, recovery and salvage tanks, bridge and fascine-layers and armoured personnel carriers. Comprehensive experience had therefore been gained in this area.

Flamethrower Tanks

Great attention was paid to flamethrower tanks. Four early versions are known, the ChT-26, ChT-130, ChT-133 and ChT-134, based on the T-26 light infantry support tank. They were used in action for the first time in the war against Finland in 1939–40. On the first day of the war against Germany, the Red Army had over 800 flamethrower tanks at its disposal.*

Between 1941 and 1945, flamethrower tanks were the most numerous Soviet specialised tanks. It is no surprise that the T-34 medium tank

---

* By comparison, in July 1941 the German armed forces had 85 Panzer II flamethrower tanks (Sd.Kfz. 122) available. The total production in 1940–2 was 155.

Command tanks were designated T-34K and T-34G. These were series-produced vehicles with a reduced magazine capacity to enable the installation of a second radio. The T-34G (G for Generals) had a transmitter with a range of 122 km.

---

provided the basis for many of these. Between January and May 1941, engineers and technicians at Factory No. 183 Kharkov worked at equipping a flamethrower tank (OP-34) from the current production. Tests proved successful and based on decisions of the Council of People's Commissars and the Central Committee of the Communist Party, the OT-34 tank with the ATO-41 flamethrower was passed for series production, the first vehicle becoming available in June 1941. Serious production was not commenced at once as a consequence of the reverses on the German-Soviet Front and the need to relocate armaments factories to the east. Contracts were placed with factories No. 174 (Machinery Factory V. I. Lenin) at Perm, later Omsk, No. 112 Krasnoye Sormovo at Ohrenburg, No. 183 Nizhny Tagil and from December 1941 the Stalingrad Tractor Factory. Between December 1941 and January 1942 the first ten OT-34 flamethrower tanks (also designated T-O34) were delivered. Decision No. 542 of the People's Commissariat for Defence of 28 July 1942 gave the go-ahead for mass production with deliveries expected to begin forthwith.

T-34/76s of various designs were converted to flamethrower tanks. Pictured here is the 1943 model from the Urals Machinery Factory, Sverdlovsk. Factory No. 222 in the Chelyabinsk region manufactured the ATO-41 flamethrower.

A characteristic feature was the ATO-41 automatic flamethrower which used powder under pressure and was delivered by Factory No. 222 at Lyuberetzky. The device was mounted in the right side of the hull with 15° traverse and −2°/+10° elevation, replacing the MG. The diameter of the nozzle was 224 mm and it could fire 16–18 bursts of flame per minute, with 105 litres of fuel available. The powder cartridges were automatically reloaded and gave a greater range compared to flamethrowers using compressed air. The range of the bursts – up to 60–65 m – could be increased to 90–100 m by the addition of a special powder (OP-2).

The crew of an OT-34/76 was three men. Outwardly the tank looked like a standard T-34 but the tanks could weigh between 26.6 and 30.9 tonnes resulting from differences in detail for various factories. Common to all were the smaller ammunition load (71–77 shells for the 76.2-mm F-34 L/41.5 main gun and 2,394–2,646 rounds for the coaxial 7.62-mm DT MG), necessary in order to create space for the fuel tank, air pressure bottles, pressure piping and the flamethrower itself. This was also the reason for the smaller crew.

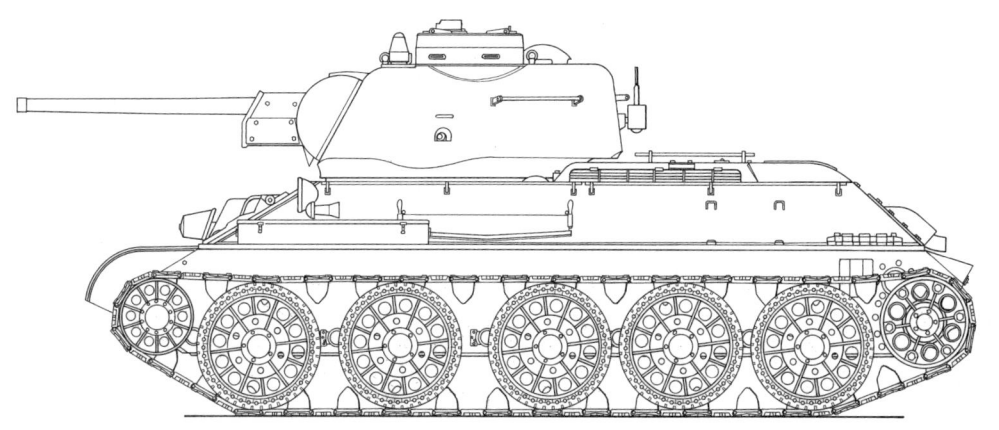
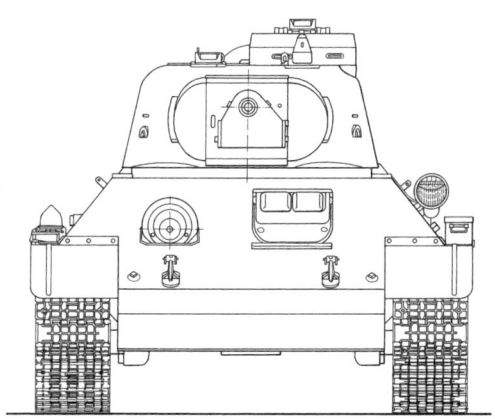

OT-34 flamethrower tank, 1943 model. (*Drawings by Robert Jurga*)

At first glance the flamethrower tank could not be distinguished from a normal tank which made it difficult for the enemy forces to pick it out as a priority target. Its disadvantages were that the flamethrower tank could only fire in the direction of travel; aiming was complicated, requiring the driver to lean to the right; noxious gases spread through the interior of the tank with every burst and affected the fighting ability of the crew; the fuel containers posed an additional danger, for if shells hit the armour in those places, flammable liquid could spread in the interior and ignite. The crew also needed fire-resistant clothing which was very uncomfortable to wear.

In 1942, Factory No. 222 produced the improved ATO-42 flamethrower designed by W. W. Kruilov which worked with a pressure of 18–21 kg/cm² (the ATO-41 had 16–18 kg/cm²). The number of bursts could be raised from 24 to 30 per minute, the range from 90 to 120 m. It was installed from 1943 in the T-34s of the current series production, and also in the OT-34/85 flamethrower tank of which Factory No. 174 Omsk delivered the first thirty by the end of 1944. The other manufacturers building medium flamethrower tanks followed in 1945. The OT-34/85 was the most advanced medium tank of its kind with typical changes: the ATO-42 at the front of the hull in place of the 7.62-mm DT MG with a crew of

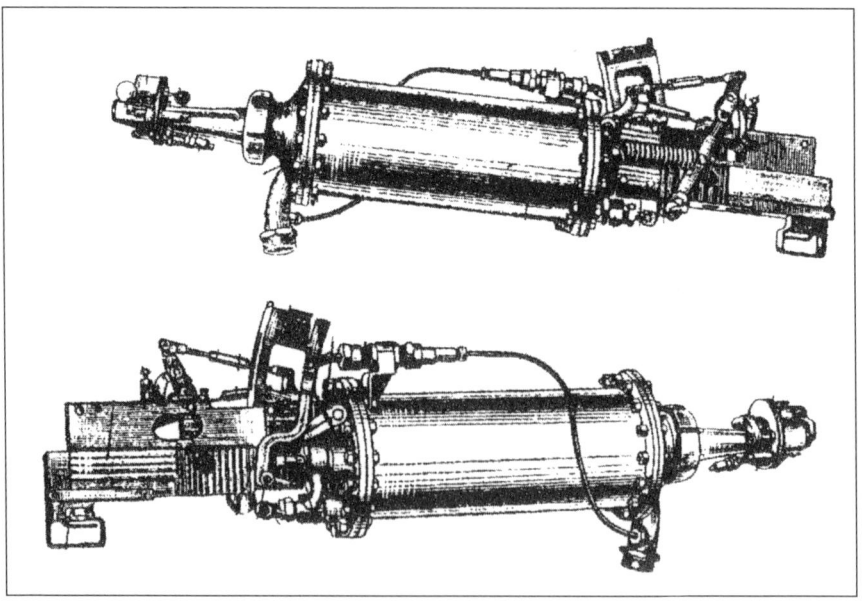

ATO-41 Automatic flamethrower.

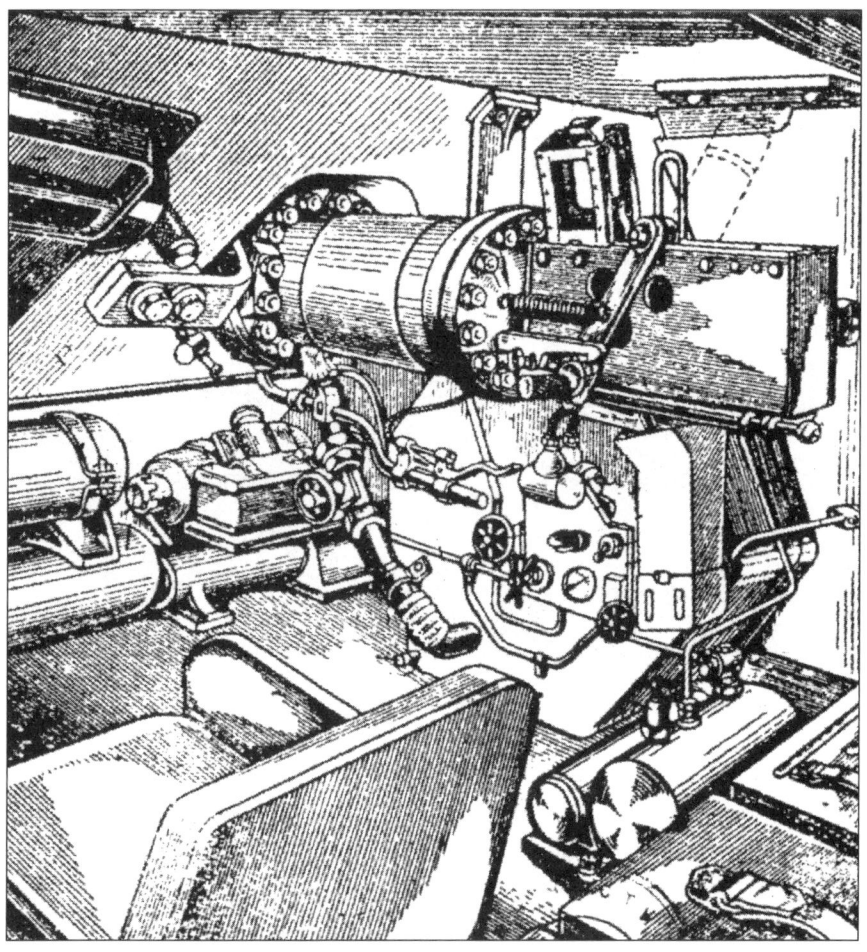

The automatic ATO-41 flamethrower (ATO-42) as installed in the T-34/76 and T-34/85 (seen from the driver's position).

four instead of five and a smaller ammunition load, but with the supply of flamethrower fuel doubled to 200 litres.

The Germans knew about the new flamethrower tank by the end of 1944. Report No. 38 by the chief intelligence officer of Foreign Armies East of 30 December 1944 states:

> Following several recent reports by prisoners of war, some of the T-34/85 tanks may have been fitted with a flamethrower in place of the gunner-radio operator and have been designated

OT-34/85 flamethrower tank with the automatic ATO-42.

T34/85 O. The substance is said to be a mixture of benzine, motor oil and a yellow powder put together in the field and can throw out streams of flame up to 100 metres in length. Further clarification is desirable.

It may be mentioned that another, very unusual but no less interesting, solution for a flamethrower T-34 tank was devised by the troops in 5th Army's sector themselves. The pioneers' FOG flamethrower with electrical ignition and 30-mm blast nozzle was used. The pressure-proof metallic cylinder with attached powder container weighed between 52 and 55 kg, and held 25 litres. Remote electrical ignition of the powder charge created a pressure of 40–50 kg/cm^2 which forced the oil from the container in bursts 50–140 m in length – depending on the type.

In workshops at the front, twenty FOG were installed on sleds each towed by a T-34 to the deployment area, the idea being that the FOG could be fired at an angle to the direction of travel. It was found that the sled restricted movement in the terrain, however. Finally two rows of five FOG each were mounted behind armour shielding above the tank's engine

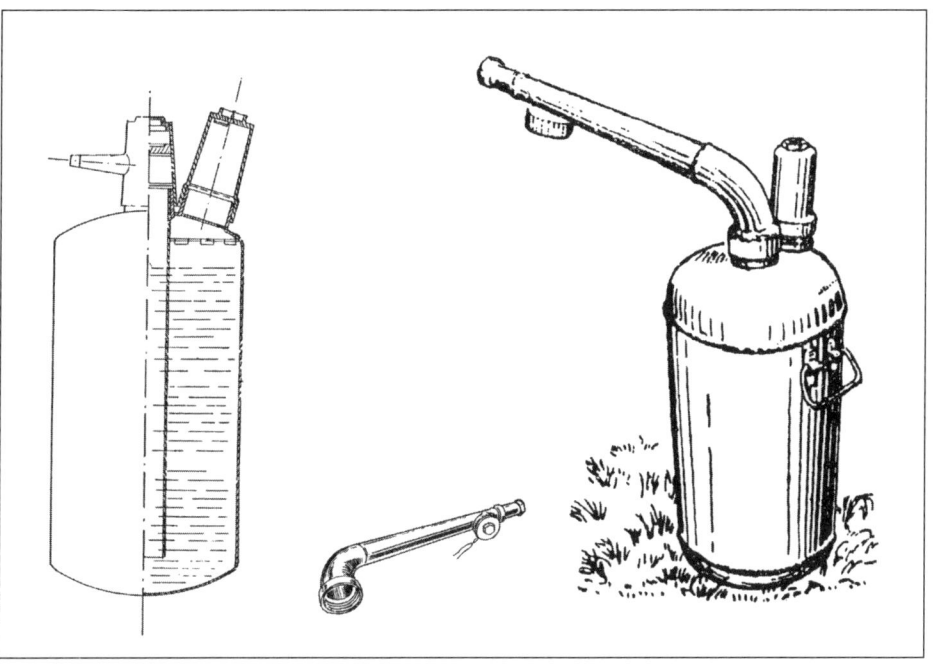

General appearance and cross-section of the FOG flamethrower with steel nozzle and powder chamber at the top. Ten of these could be installed in an armoured battery on the engine compartment cover of a T-34/76.

A T-34/76 fitted with the FOG equipment.

OT-34/76 and OT-34/85 Flamethrower Tanks Completed, 1941–45

	Stalingrad Tractor Factory	Factory No. 183 Nizhny Tagil		Factory No. 112 Krasnoye Sormovo		Factory No. 174 Omsk		Totals
	OT-34/76	OT-34/76	OT-34/85	OT-34/76	OT-34/85	OT-34/76	OT-34/85	
1941	–	–	–	–	–	–	–	–
1942	31	172	–	106	–	–	309	618
1943	–	90	–	229[1]	–	159[2]	–	478
1944	–	–	–	52	–	331	30	413
1945	–	–	91	–	55[3]	–	155	301
Totals	31	262	91	387	55	490	494	1,810

Notes: 1. Seventy-seven radio-equipped; 2. One with ATO-41, the remainder with ATO-42. Seventy-one radio-equipped; 3. All with radio equipment.

compartment, a PT-22 ignition machine being fitted additionally. This makeshift solution was adopted.

The Red Army had formed its first flamethrower units in the mid-1930s but did not have the OT-34 flamethrower tank until the late summer of 1942. That August, three flamethrower tank battalions (Nos. 502, 503 and 507) were formed, each with five KV-8 heavy flamethrower tanks, fourteen OT-34s and two T-34s at Battalion HQ, later increased to five KV-8s, sixteen OT-34s and ten T-34s equipped with radios. In October 1942 a flamethrower tank brigade was set up comprised of an HQ, HQ Company, two flamethrower tank battalions and a technical security company (total inventory thirty-six KV-8s, eighteen OT-34s and five T-34s). On 1 January 1943, besides the KV-8s, there were now 309 OT-34s available and a year later 787. Red Army flamethrower tank units played an important role in the closing phase of the Second World War in actions against fortified positions and in towns identified as fortresses or defended areas.

Mine Clearance Tanks

The Red Army's initial work on mine clearance dated from 1932 when light and medium tanks fitted with blades, rollers and mine-impact devices, later explosive devices, had the objective of detecting and overcoming obstacles, booby traps, mine barriers and minefields laid in the open or dug in by enemy engineers. Nothing is known of any series production although at the outbreak of war experience in mine clearance was available.

1944 model T-34/85 mine-clearance tank with PT-3 equipment fitted.

The contract for the development of armoured mine clearance equipment for the T-34 and KV-1 was issued by the People's Commissariat for Defence at the beginning of 1942. Colonel P. M. Mugalev played a key role in the development work. He had been engaged previously in designing a road-building machine as a research assistant at the Academy of Pioneer Troops. As he stated later, it was during the Winter War with Finland in 1939–40 that he saw the devastating effects of mines with his own eyes. This experience was not without influence in his later work developing the 'Traltshik', the PT-3 anti-mine roller capable of attachment to a T-34 or KV-1 at the front of the hull by the use of a holding harness with frame and steel hawser. The device weighed 5.3 tonnes, was 2.78 m long and 3.6 m wide overall. On each side were arranged five revolving steel discs with a diameter of 1.1 m diameter. By their weight the rolling segments set off anti-personnel and anti-tank mines over a broad swathe, both in the open or dug in. The speed of sweeping was 8 km/h. To reach operational areas the roller segments could be dismantled for carriage individually on ZIS-5 or Studebaker trucks. It took fifteen minutes to refit them to the tank. The first PT-3 equipment was delivered at the end of 1942 to the 4th Independent Guards Tank Regiment on the Voronezh

1944 model T-34/85 mine-clearance tank on the move. When not in use the rollers were dismounted and carried aboard a truck.

Front for trials. There were many doubters but their criticisms led to improvements.

In June 1943 a beginning was made by creating tank pioneer regiments with twenty-two T-34/76 and eighteen PT-3 mine clearance units (also known as Special Regiments with Landmine Clearance Equipment). In 1944 the T-34/76 was replaced by the T-34/85 and later the new T-44. Anti-mine tanks fitted with this equipment proved their worth on all fronts; they underwent further development after the war and were copied abroad.

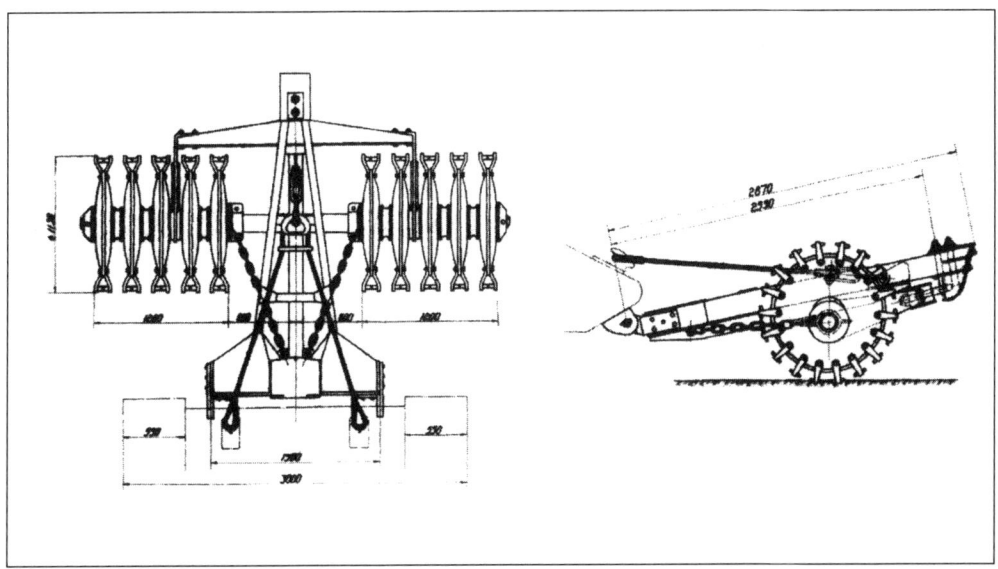

The PT-3 mine-clearance equipment was of the roller type and weighed 5.3 tonnes.

Recovery and Bridgelaying Tanks

Officially there were no T-34 recovery tanks. None were diverted for the purpose from series production, but instead troops converted damaged

T-34-T recovery tanks (tank towing machines) were former combat tanks repaired and converted. They had two crew; the equipment included a 25 m-long steel hawser.

A T-34-T during the arrival of the Red Army in Prague on 9 May 1945.

T-34s in the field. Several kinds of lifting crane are known, with A-frame or lattice booms, some with pulleys, others with electrically-driven winches having a lifting power of one tonne.

In 1942 a bridge-laying tank, designated TM-34, was tried out on the Leningrad Front, manufacturer being the tank repair factory No. 27. Other special vehicles such as T-34 fascine carriers, were devised at the Front.

Apart from their role as flamethrowers and mine-clearance vehicles, T-34s were only diverted from the current series production in large numbers as command tanks for companies, battalions and regiments, some even for generals. These could be distinguished by their comprehensive array of communications equipment.

The AT-42 transport tug (Object 42) based on the T-34/76, did not progress beyond the project stage. In 1944 a further design based on the

T-34/85 chassis designated AT-45 was to be equipped with a throttled 350-hp W-2 diesel. This development was incorporated into the T-44.

The T-34 for Active Propaganda

In 1942 on the Bryansk and North-West Fronts, T-34/76s equipped with twin loudspeakers were used to broadcast propaganda to German troops. Dependent on the weather they could be heard from about 2 km away.

1942 model T34/76, as loudspeaker tank for active propaganda. Photo taken in October 1944.

4 Self-Propelled Guns

'*Samochodnyie ustanovki*' = SU, in English 'self-propelled chassis' or SP gun, was the abbreviation used for Russian vehicles on which a gun, howitzer or recoilless gun could be mounted. The chassis could be tracked, wheeled or a half-track. In the Second World War they had no turret, but a fixed casemate. The advantages of an SU gun over a tank included its heavier calibre of main weapon, lower profile and comparatively light weight. The tasks it could be asked to handle were against targets offering serious resistance, including German panzers. The Soviet Union had considerable experience in the design of SU chassis. Up to 1940, over twenty projects, prototypes and series-completed vehicles of this kind are known based on adapting medium and heavy tank designs.

Project U-34 in 1942 – already prefiguring the form of the later SU-122, -85 and -100 designs.

The SU-122. The 122-mm M-30 M.1938 L/22.7 howitzer could effectively pin down and wipe out enemy troops even if they were in cover. Armoured vehicles could be taken under direct or indirect fire.

SU-122

The first project based on the T-34 commenced in 1942. The responsible engineers at the design bureau of the Urals Heavy Machinery Factory 'Ordzhonikidze'/Sverdlovsk (Uralmash, USTM) were N. W. Kurinuim and G. F. Ksyuninui. The project was designated U-34. It had a fixed armoured superstructure sloping at the front and sides and a 76.2-mm gun (traverse 20°, elevation −4°/+31°). Apart from being lighter than the T-34 by two tonnes and its overall height of 2 m it had no advantages, for the aim was to increase firepower. In the early months of 1942, the Committee for Artillery Armaments wanted an SP chassis with 45-mm armour for the 122-mm M-30 M.1938 L/22.7 howitzer, a gun already in mass production.

F. F. Petrov's design bureau at Factory No. 9 worked in collaboration with the Sverdlovsk factory on Project U-35. The results were presented in August of that year and in many respects were seen to resemble Project U-34 with its fixed superstructure, the characteristic forward-sloping gun shield providing space for the five-man crew and the howitzer with recoil

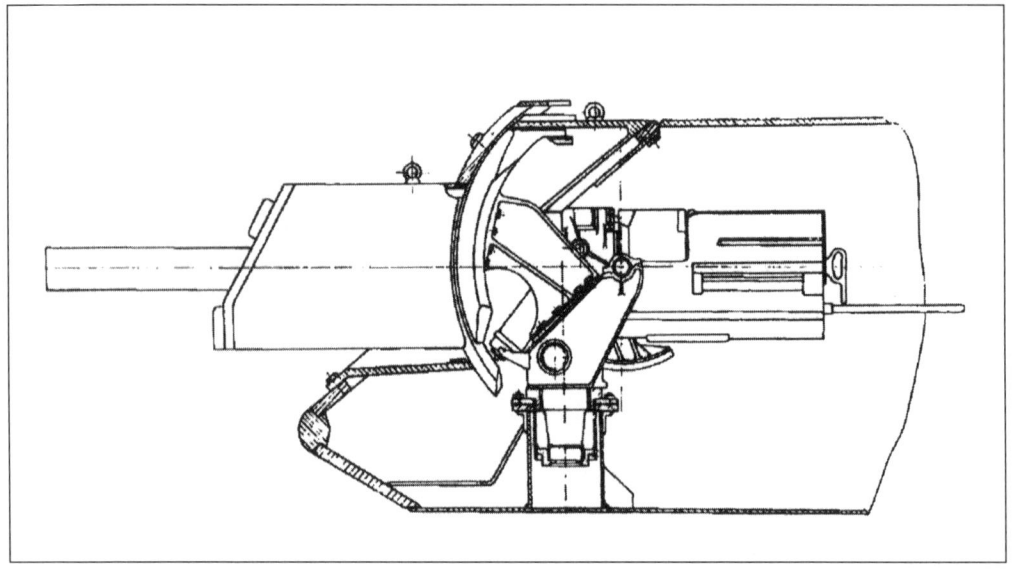

The mounting of the 122-mm M-30 M.1938 L/22.7 howitzer, showing the upper body fixed on the floor of the hull.

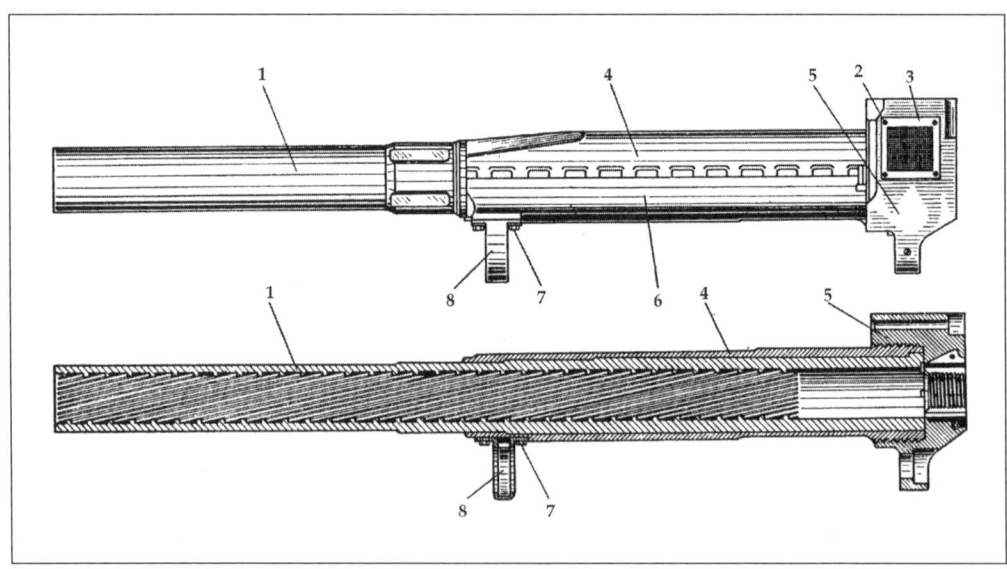

Barrel of the 122-mm howitzer at the Kubinka Tank Museum:
Key: 1. Barrel lining; 2. Screw; 3. Diagram; 4. Barrel jacket; 5. Baseplate; 6. Sliding surface, left; 7. Screw; 8. Location of brake cylinder.

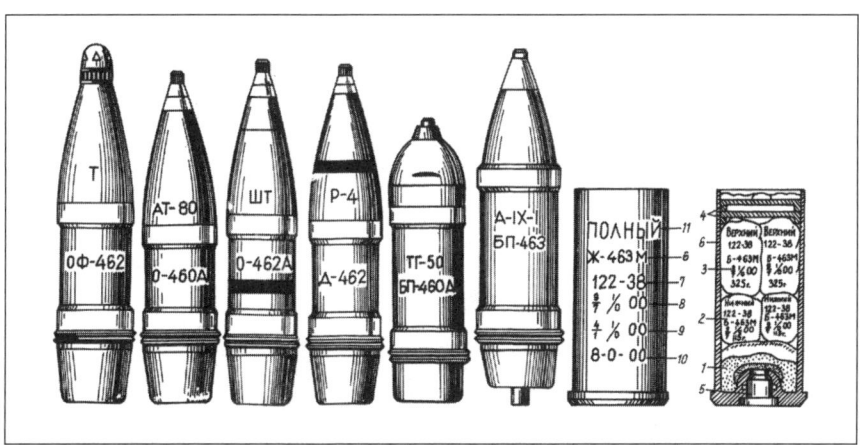

Ammunition for the 122mm M-30 M.1938 L/22.7 howitzer:
(From left) OF-462 fragmentation shell (weight 21.76 kg); OF-460 cast steel fragmentation shell (weight 21.76 kg); OF-462 cast steel fragmentation shell (21.76 kg); D-462 smoke shell (22.55 kg); BP 460A hollow-charge shell (13.34 kg); BP 463 hollow-charge shell (14.8 kg); Cartridge case.

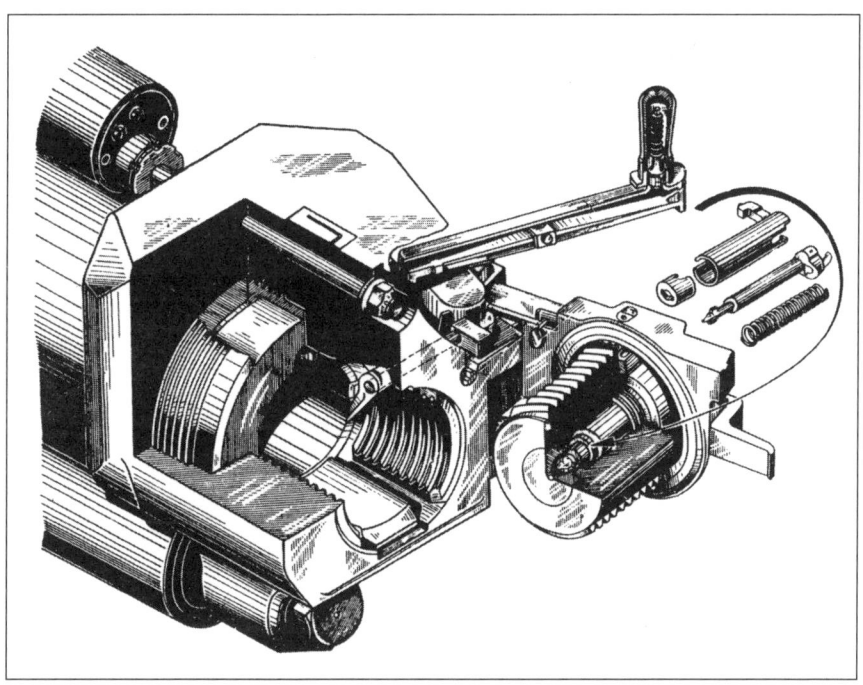

The screw-piston breech of the SU-122's M-30 M.1938 L/22.7 howitzer.

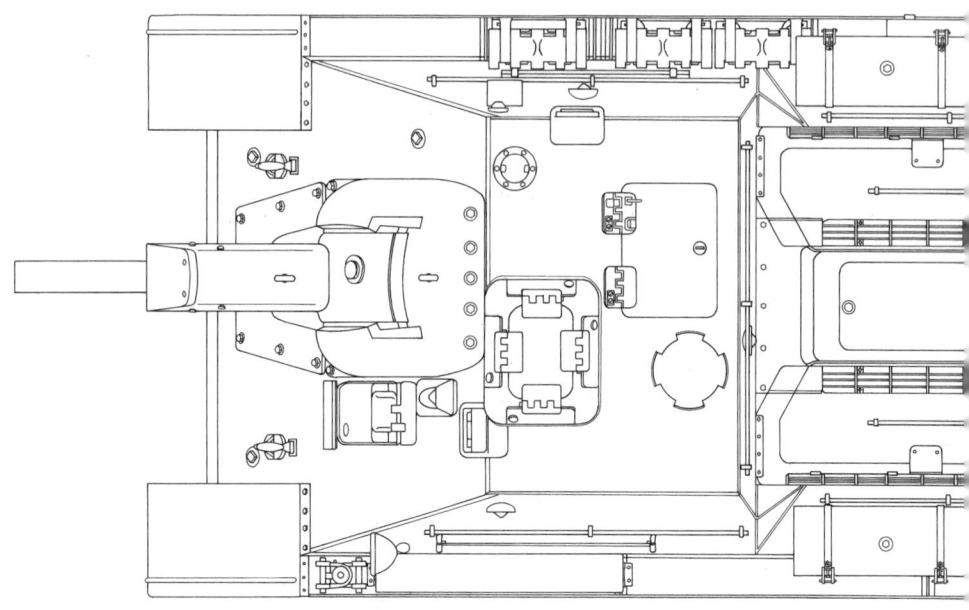
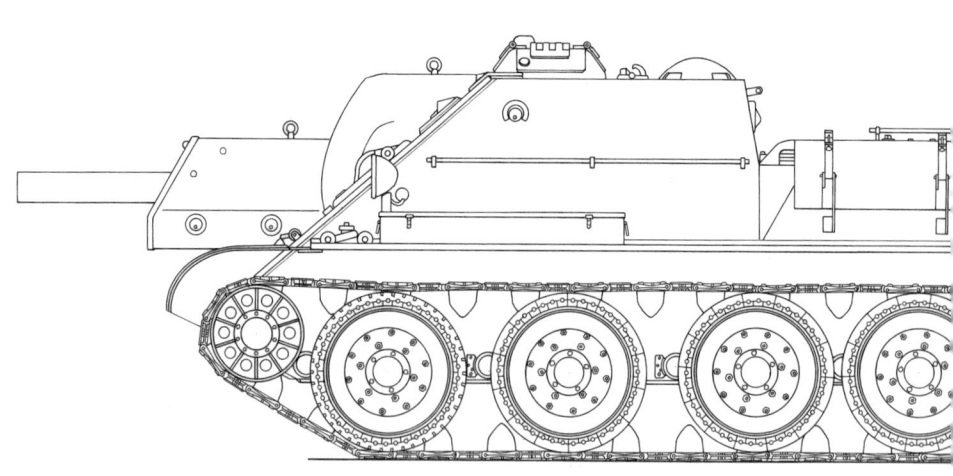

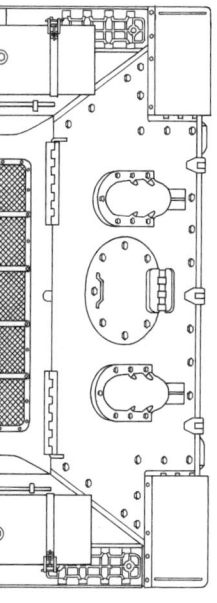
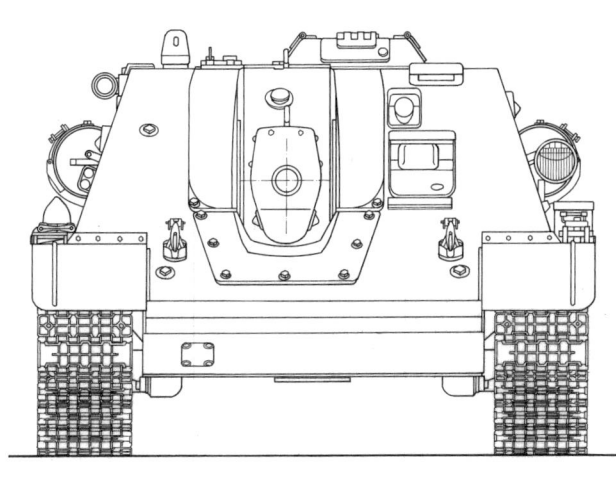

The SU-122. (*Drawings by Robert Jurga*)

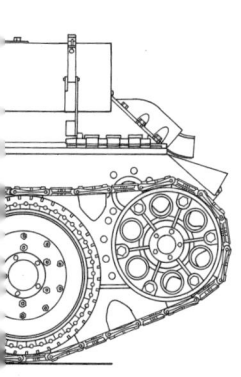
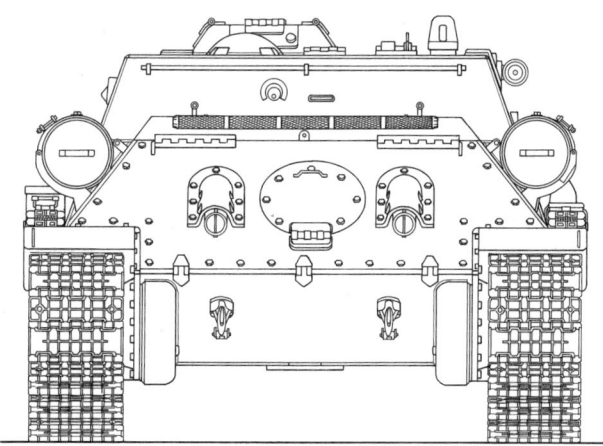

An SU-122 captured by a German SP-gun unit during the Battle of Kursk, July 1943.

brake and recuperator. This latter was situated on the hull floor of the upper body with the traverse and elevation machinery (20° to either side, -43°/+25° elevation range). A sight and PG-1 or PG-1 M panoramic telescope with collimator was installed. The maximum range for indirect fire was given as 8,000 m, rate of fire two rounds per minute. Forty rounds were carried. The ammunition had to be put together before loading – first the projectile, then the cartridge (with propellant and one to six supplementary charges). OF-462 and O-460 heavy high explosive/fragmentation shells weighing 21.76 kg were fired, also smoke shells and starshell. For engaging panzers BP-460 hollow-charge shells (muzzle velocity 335 m/sec) were available, and later the BP-463 (muzzle velocity 460 m/sec), which weighed 13.34 kg and 14.8 kg respectively. The optimum range in combat with panzers was 400–600 m. Ready for action the SU-122 weighed 29.6 tonnes, overall height being 2.235 m.

On 19 October 1942 the People's Commissariat for Defence followed the recommendation of the Artillery Committee and the SP chassis was officially approved. The fitting of the 122-mm U-11 L/22.7 howitzer into the KV heavy tank (designated as KV-9) was finally rejected after Factory

Trials of a captured SP gun by Weapons Testing Unit 6/Army Weapons Office at Sankt Johann, Tyrol, on 9 March 1944.

Image of the SU-122 from a tank recognition table. The type was known to the Germans as the 'Sturmgeschütz T-34'.

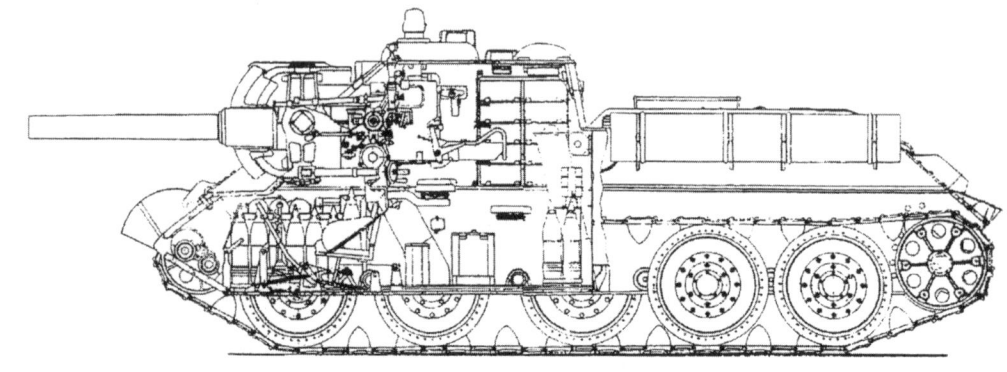

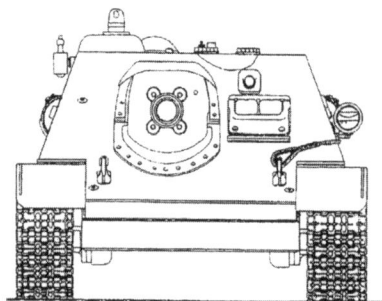

Cross-section of the SU-122 III design.

SU-122 with the 122-mm D-11 howitzer.

The SU-122 III with a five-man crew weighed 30.1 tonnes. The magazine held forty rounds for the 122-mm D-5 howitzer.

No. 9 at Sverdlovsk had already delivered ten units in April 1942. The plan was to deliver the first twenty-five T-34-derived SU-122s from December, 100 in the first quarter of 1943 and from April 1943 125 vehicles a month. In the event, Factories No. 592 and Uralmash supplied two dozen by the end of the year and, until the cancellation of production in the late summer, a total of over 600 of the type.

The first SP-gun regiments received seventeen light (SU-76) and eight medium (SU-122) vehicles: from April 1943 regiments were formed and supplied with sixteen SU-122s only, the authorised strength later being raised to twenty.

The Germans, planning a 12.2-cm SP assault gun, assessed the SU-122 as 'a useful tank, especially against personnel. For engaging panzers, on account of its short barrel length it is only suitable at short range; difficult to attack frontally, easier from the sides.'

In the spring of 1943, contracts were issued for the development of the SU-85 SP gun. The 85-mm gun was to be installed on gimbal mountings providing a better solution than the 122-mm M.1938 howitzer. Two new howitzers were then developed specially for use in SP guns. In July

1943 the 122-mm D-6 howitzer was fitted in an SU-122 III which then weighed 30.1 tonnes. The D-11 (U-11) for a modernised version, the SU-122M, was presented in the autumn of 1943. At 31.4 tonnes it was comparatively heavy and could only make a top speed of 47 km/h. Both were fitted with the new T-10 panoramic telescope. Neither type went into series production.

The SU-122 was undoubtedly a successful design but even so production came to an end in August 1943. The 122-mm howitzer lacked the capability to engage German panzers and SP guns at long range, and it had become urgently necessary for new designs to overcome the growing inferiority of Russian tanks. To increase fighting value would take time, and now the yardstick for new tanks and SP guns was the German Tiger (Sd.Kfz. 181) which had to be destroyed before it got within 1,000 m.

SU-85

In April 1943 the Artillery Committee of the Red Army put together the tactical and technical requirements for an SP-chassis with an 85-mm gun. It was hoped to reach the necessary level for modernising the SU-122 (SU-122M). On 5 May 1943 the State Defence Committee made a decision regarding the development of new tanks and SP guns to include the SU-85, the gun designs of which were to be undertaken by three agencies working separately. The three prototypes available that month differed as regards the gun but were outwardly distinguishable by their gun shields.

- SU-85 I with 85-mm S-18-1 gun (Central Artillery Design Bureau and Uralmash).
- SU-85 II with 85-mm D-5-S-85 gun (Factories No. 9 and 50).
- SU-85 IV with 85-mm S-18 gun (Central Artillery Design Bureau).

All three were tested at the Gorochovetz firing grounds where the decision fell in favour of the SU-85 II even though at 1,370 kg its gun was considerably heavier.

On 7 August its introduction was confirmed under the official designation SU-85. At this time the first four series vehicles had already been delivered; another 2,650 (amongst them 315 SU-85Ms) were to follow by November 1944 when they were supplied to the medium SP gun regiments which had been reorganised at the beginning of 1944 and had

Three prototypes of a new SP gun were presented in May 1943. The decision was made in favour of the SU-85 II with the 85-mm D-5-S-85 L/51.6 gun. At this time (August 1943) the first four series-produced SU-85s had already been delivered.

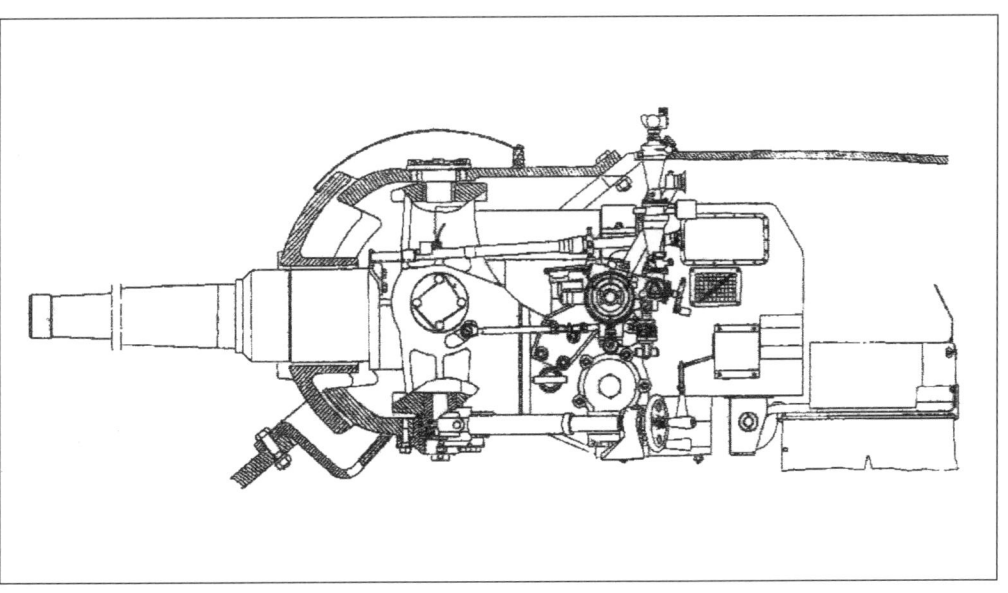

Mount for the 85-mm D-5-S-85 L/51.6 gun on the SU-85.

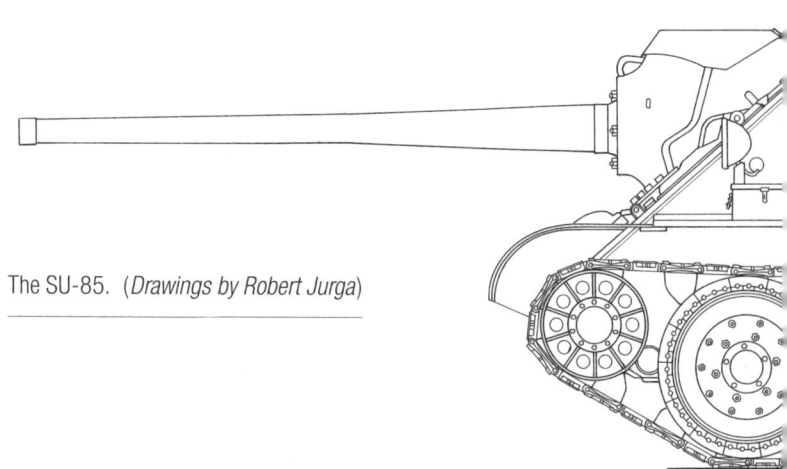

The SU-85. (*Drawings by Robert Jurga*)

Overhead view of the SU-85 showing the casemate superstructure with rounded gun shield and the arrangement of hatches and periscopes.

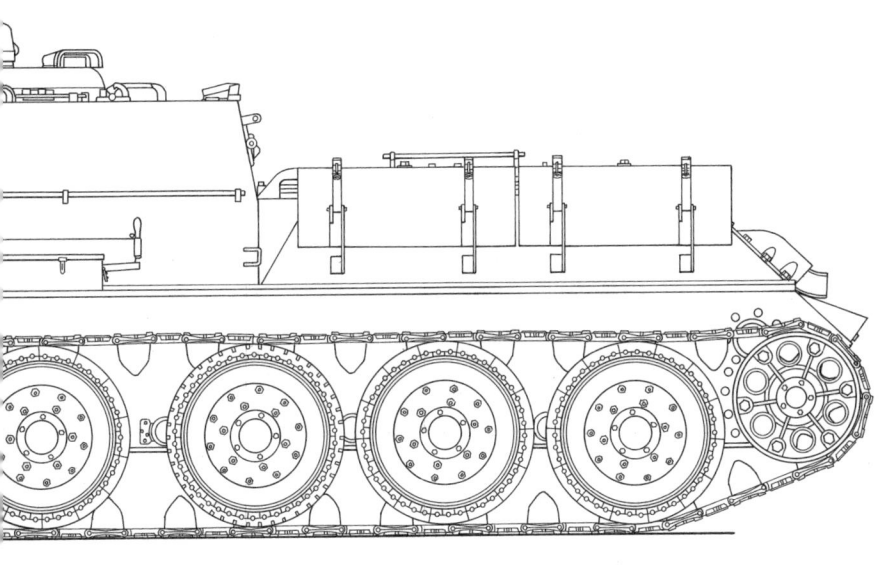
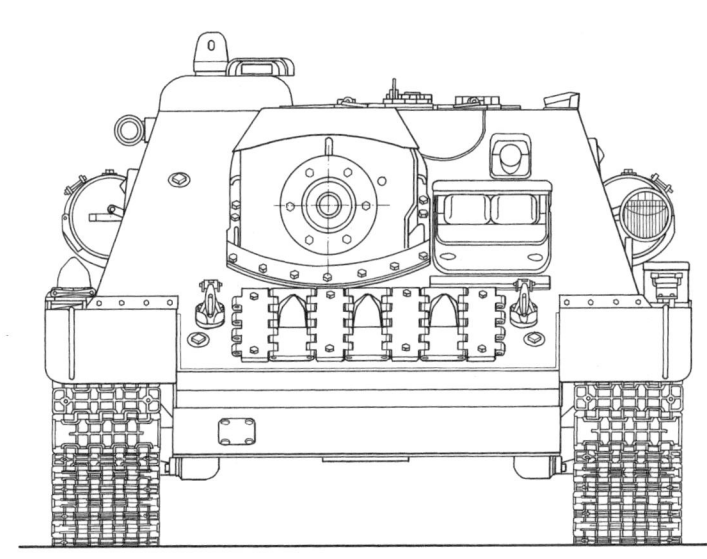

The SU-85. The armour was of pressed steel with welded seams, was well sloped and offered good protection.

more than 21 SU-85s each. In addition, the SP-gun brigades of the six tank armies also received them.

The basic structure of the SU-85 resembled that of the SU-122; chassis, engine and drive unit were as for the T-34 tank, at the front there was a fixed armoured structure with the driver's and fighting compartments. The 85-mm D-5-S-85 L/51.6 gun had a gimbal mounting with frame, barrel cradle and shield of 52-mm cast steel (traverse either side 20°, elevation –5°/+25°). The 10-T-15 periscope (× 2.5 and 15°) served for aiming purposes, replaced later by the TSch-15 (× 4 and 16°). The four-man crew was located as follows: (i) the commander to the right of the gun below a square cupola with periscope, and no exit hatch; (ii) the gunner seated to the left of the gun; (iii) the loader behind him; and (iv) the driver seated in the armoured hull forward, on the left side. In the ceiling of the armoured superstructure, near the driver's hatch with two periscopes, were two more hatches and another five periscopes. The crew also had an emergency escape hatch in the hull floor forward to the right.

Outwardly the SU-85M differed from the tanks of earlier runs by having the commander's cupola equipped with five viewing slits, a rotatable periscope and a two-part hatch (as on the T-34/85). The gunner's hatch had

Both above: An SU-85 destroyed in March/April 1944 by a German StuG III 7.5-cm SP gun (Sd.Kfz. 142.1) at Chertkov.

Above: An SU-85 put out of action by a hit to the right flank.

Left: An SU-85 during the battle for Berlin, end of April 1945.

Above & top right: The 85-mm D-5-S-85 L/51.6 gun in its frame.

Rear view of the SU-85 showing the typical characteristics of the 1943 model T-34. In the summer of 1944 an SP regiment consisted of four batteries of four SU-85s each. The regimental HQ company had a T-34 command tank.

An SU-85M, recognisable by the commander's cupola on the right side of the vehicle. Production of this type ceased in November 1944. Photograph taken in Germany, May 1945.

Armour Penetration, SP Guns on T-34 Chassis							
Range	122-mm M-30 L/22.7 howitzer		85-mm D-5-S-85 L/51.6 gun			100-mm D-10-S L/56 gun	
	Hollow-charge shell		AP shot			AP shot	
	BP-4631[1]	BP-460 A2[2]	BR-365	BR-365 K	BR-365 P3[3]	BR-412	BR-412 B
500 m	–	–	111	101	143	155	162
1,000 m	–	–	102	84	103	135	149
1,500 m	–	–	93	69	–	116	132
2,000 m	c. 90–100[4]	c. 100[4]	85	57	–	99	124

Notes: Notional angle of impact 90° in all cases. 1. Maximum range 4,000 m, against armour up to 660 m. 2. Maximum range 2,000 m, against armour 400 m. 3. Sub-calibre round. 4. Hollow-charge penetration unaffected by range.

a different design. Another feature of note was the two ventilators with armoured covers behind the commander's cupola. The frontal armour was now 75 mm, and the weight of the SU-85 had risen to 31 tonnes.

Forty-eight fixed rounds were carried, nineteen of them being BR 365 or 365 K AP shells (weight 9.2 kg and 9.34 kg respectively, muzzle velocity 800 m/sec), nine BR 365 P heavy sub-calibre AP shells (5.35 kg, muzzle velocity 1,050 m/sec) and twenty O.365 K fragmentation shells (weight 9.54 kg, muzzle velocity 793 m/sec). The AP shells could penetrate 102 mm and 84 mm of armour respectively at 1,000 m, while the sub-calibre shells had a penetration of 103 mm at 500 m. The rate of fire was six to seven rounds per minute.

The German reports described the SU-85 as an '8.5-cm gun on a T-34 chassis' and as a 'support weapon especially useful for anti-panzer work. From ahead difficult, from the sides easier to engage.' In 1944 the Germans expected to deploy more powerful armoured fighting vehicles – reason enough not to allow any interruption to the efforts to increase the firepower of the SU-85. The Sverdlovsk designers worked at this task from August 1943. The SU-85 BM-I and BM-II mounted the 85-mm D-5-S-85 BM gun, developed at Factory No. 9 under specifications requiring the ability to penetrate the 100-mm front armour of the Tiger from 1,500 m (muzzle velocity 800–900 m/sec). For the SU-85 BM II, tested in February 1944, the military was hoping for a further increase in performance to a range of 2,000 m, and in order to increase the muzzle velocity of the AP shell to 1,015 m/sec, the gun was given an additional smoothbore section.

The SU-100. The enormous overhang of the barrel is striking.

SU-100

The work to increase the fighting effectiveness of the SU-85 was discontinued at the beginning of 1944 when the decision was taken to fit the more powerful 100-mm D-10-S L/56 gun (Object 138) on a T-34 chassis. The well-proven design bureau of the Urals Machinery Works at Sverdlovsk was chosen for the chassis while the gun was to be supplied by F. F. Petrov's bureau at Factory No. 9, Sverdlovsk. The SP gun received the designation SU-100 and was tested in February 1944 on the firing range at Gorochovetzk. Apart from the armament, there were only minor differences from the SU-85M. The front armour was 75 mm, a 75-mm thick calotte of cast steel was placed ahead of the 40 mm thick armoured shield of the gun (traverse either side 16°, elevation −3°/+20°). The mounting with frame, barrel cradle, shield and calotte was therefore modified. For targeting, a TSch-19 telescopic sight (× 4 and 15°) and a panoramic

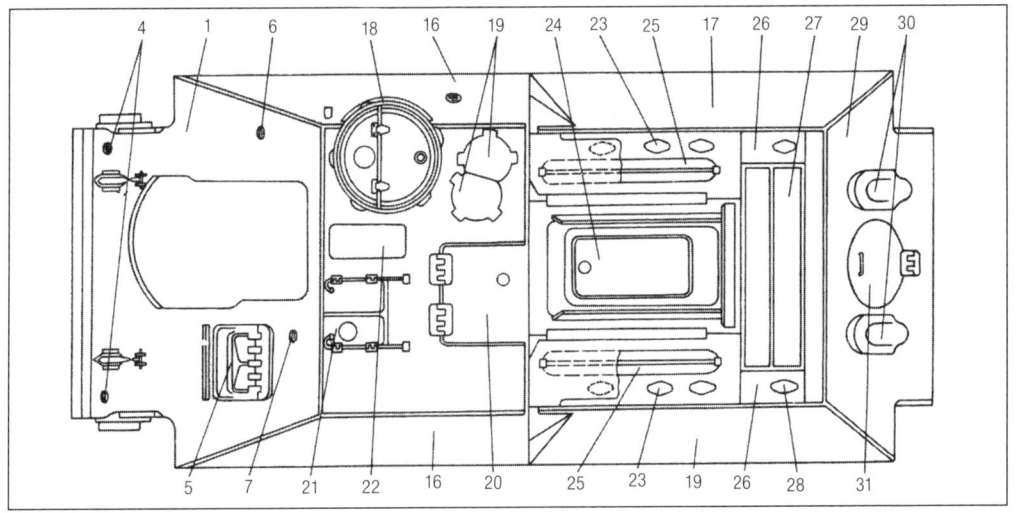

Hull of the SU-100:

Key (both illustrations): 1. Upper front armour; 2. Lower front armour; 3. Crossbeam; 4. Locking screw for track tension device; 5. Driver's hatch cover; 6. Locking screw, refuelling point; 7. Pistol port; 8. Lower side armour; 9. Thrust bearing; 10. Housing; 11. Bore-hole for swing-arm axles; 12. Swing-arm limiter; 13. Inlet for swing-arm pivot; 14. Abutment of swing-arms; 15. Track pin deflector; 16. Side armour, forward; 17. Side armour, rear; 18. Commander's cupola; 19. Protective caps for ventilator motors; 20. Access hatch cover; 21. Cover for panoramic telescope hatch; 22. Armour cap; 23. Fuel intake cover; 24. Cover for hatch below motor; 25. Louvres for air intake; 26. Armour plating over fuel tanks; 27. Louvre, rear; 28. Refuelling cover, rear tank; 29. Rear armour, upper; 30. Caps for exhaust tubes; 31. Transmission hatch cover; 32. Fighting compartment armour; 33. Radiator louvre protection; 34. Opening for aerial; 35. Pistol port; 36. Gunnery room, rear armour; 37. Rear armour, centre; 38. Lower rear armour.

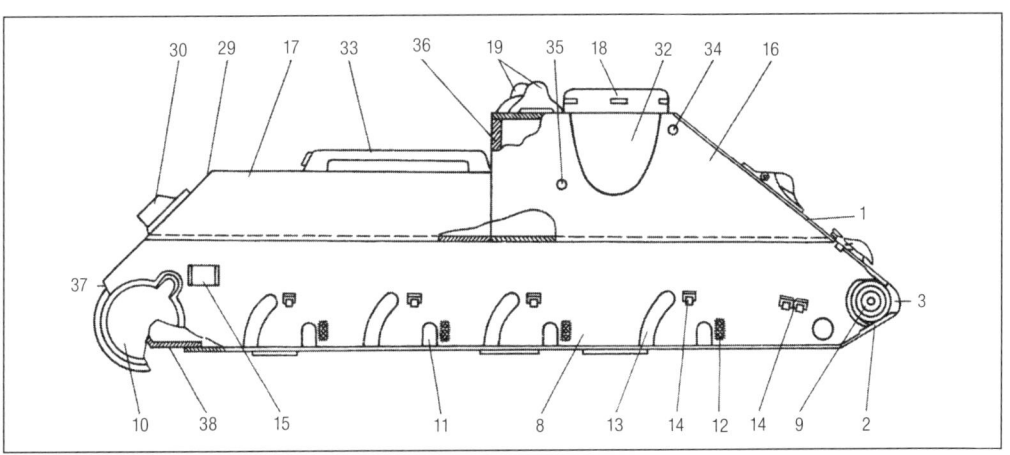

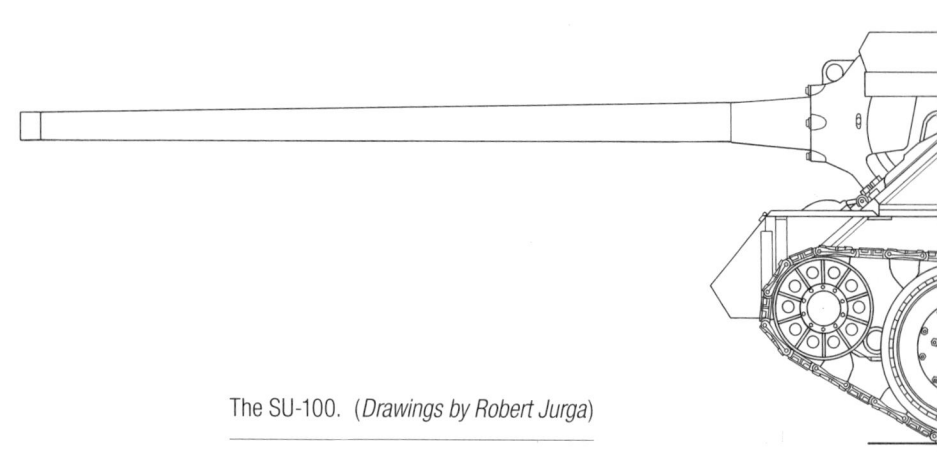

The SU-100. (*Drawings by Robert Jurga*)

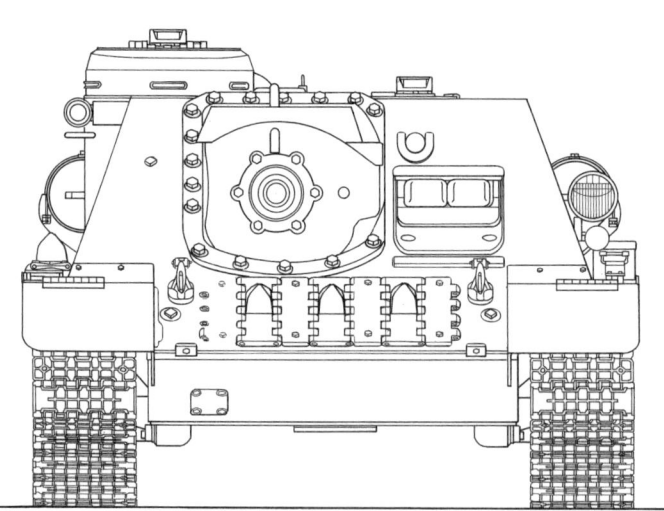

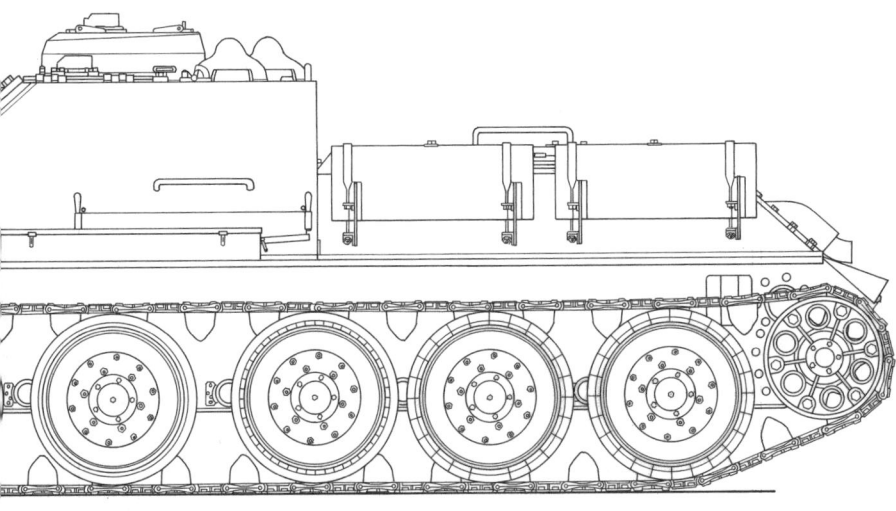

Fighting compartment in SU-100 with the 100-mm D-10-S L/56 gun.

The SU-100 with the 100mm gun D-10-S L/56 gun was delivered from September 1944.

telescope (× 3.7 and 10°) were fitted. The heavier armour and gun caused an increase in weight to 31.6 tonnes and a reduction in the top speed to 50 km/h. The 100-mm D-10-S M.1944 L/56 gun had substantially greater firepower using 15.8-kg AP shells (muzzle velocity 897 m/sec) and 15.9-kg O-412 fragmentation shells (muzzle velocity 892 m/sec). The maximum range was 15,400 m. So-called UOF-412 'naval fragmentation shells' were also available. (See p. 183 for performance of AP ammunition.) This level of performance meant that Tigers and Panthers could now be engaged effectively at long range.

Series production began in September 1944, and by December the type had replaced the SU-85M. Around 3,000 SU-100s were delivered until

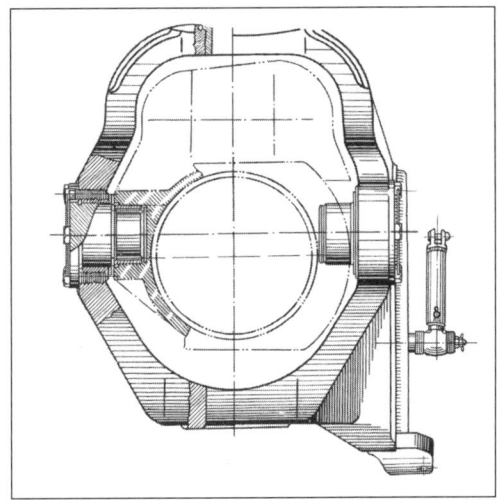

Above: Frame for the 100-mm gun (seen from the front). It is connected above and below with vertical spigots to the armoured housing.

Below: Cross-section of the 100-mm D-10-S L/56 gun.

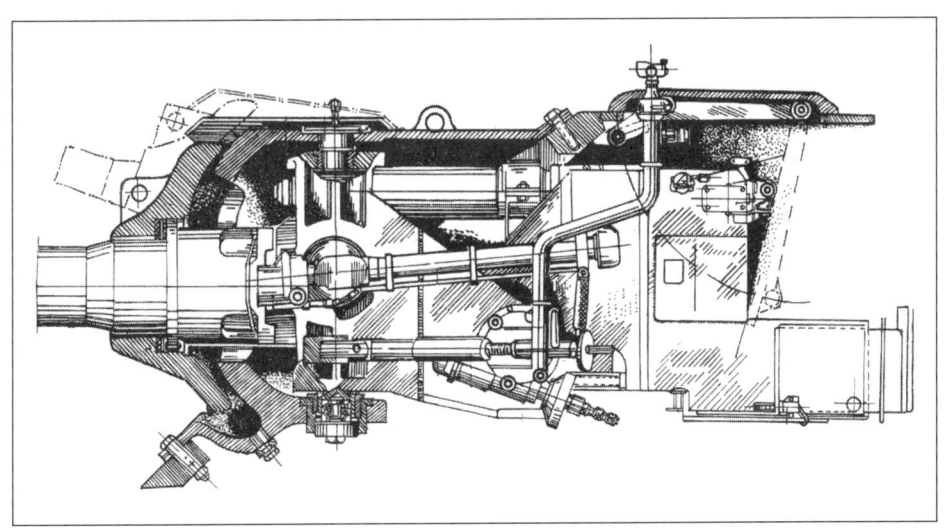

SP Guns on T-34 Chassis Completed 1942–46				
Year	SU-122	SU-85	SU-100	Totals
1942	26	–	–	26
1943	611	761	–	1,372
1944	–	1,893	500	2,393
1945	–	–	2,285	2,285
1946	–	–	252	252
Totals	637	2,654	3,037	6,328

the end of production in March 1946. On 1 July 1945 the stock of these vehicles stood at 1,560, most of them with medium SP gun regiments and brigades.

There were also attempts to arm the SU-100 with an even more powerful gun. In September 1944, the SU-100P was armed with the 122-mm D-25-S L/47.1 gun (traverse 15° either side, elevation -2°/+20°) and a TSch 17 (× 4 and 16°) telescopic sight. The gun could fire the 25-kg BR 471 AP shell to penetrate 152 mm armour from 1,000 m, and 122 mm from 2,000 m. In the zone of major clashes with enemy panzers – 1,000–1,500 m – the 100-mm gun provided no advantage, but at 2,000 m was superior; however, the greater weight of the vehicle (32.8 tonnes) and slower rate of fire (two to three rounds per minute) rendered the SP-100 inferior. The ammunition had to be put together before loading in the breech, and another disadvantage was the enormous overhang of the barrel which had increased from 2.03 m (SU-85) to 3.52 m (SU-100) and was now 3.3 m. This limited the tactical mobility over rough terrain and in woodlands as well as urban areas and was also a problem for railway transport.

In their plans for a successor, the designers took this experience into account. In autumn 1944 at Sverdlovsk they worked on the SU-101, at first on the basis of the T-34/85 chassis, later on that of the new T-44 medium tank. The layout of the interior was revolutionary with drive unit forward, then came the driver's compartment, the W-44 diesel installed crosswise, and behind it the fighting compartment with the 100-mm D-10-5 L/56 gun (traverse 22.5° either side, elevation -2°/+20°). A TSch 19 (× 4 and 15°) telescopic sight was used and a panoramic telescope was also fitted. Although the gun was 5.606 m long overall, it extended beyond the hull by only 76.5 cm (on the SU-102 by 89.5 cm).

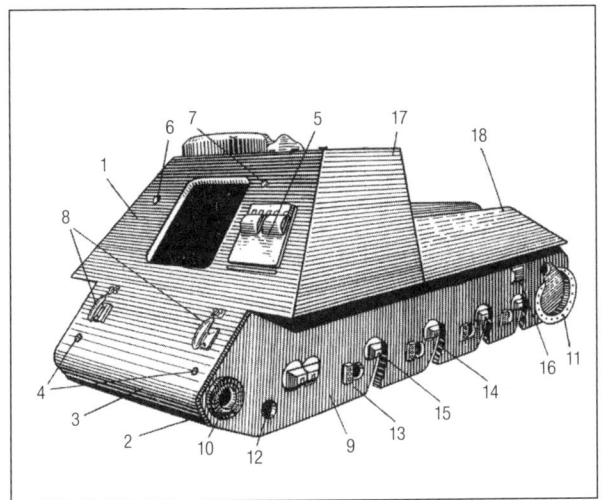

Hull components, SU-100:
Key: 1. Upper front armour; 2. Lower front armour; 3. Hull cross-beam; 4. Screws for track tension device; 5. Driver's hatch; 6. Oil filler, forward oil tank; 7. Pistol port; 8. Towing hooks; 9. Lower vertical side armour; 11. Thrust bearing of axle; 12. Housing for lateral countershaft; 13. Hole for the swing-arm axle bearing of the leading road wheel; 14. Swing-arm limiter; 15. Entry point for the pivot of the swing-arm; 16. Swing-arm support; 17. Track bolt deflector; 18. Forward lateral armour; 19. Rear lateral armour.

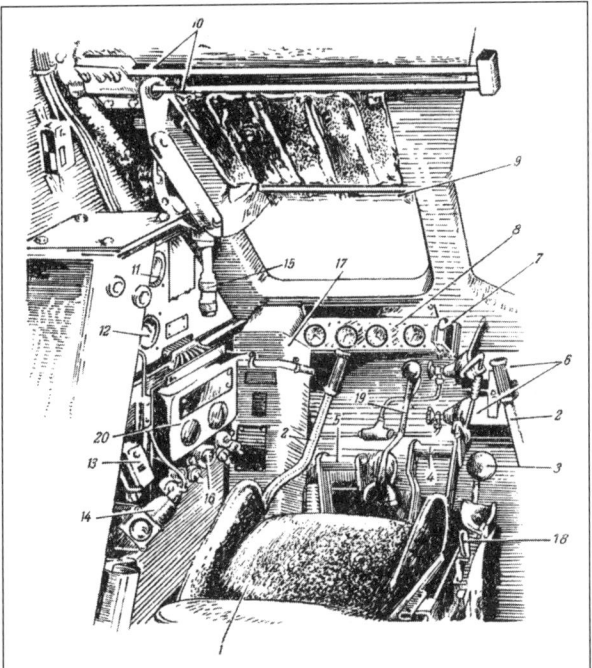

Driver's compartment, SU-100:
Key: 1. Driver's seat; 2. Steering levers; 3. Foot lever; 4. Brake pedal; 5. Clutch pedal; 6. Compressed-air bottles for cold engine start; 7. Instrument panel lamp; 8. Instrument panel; 9. Periscope; 10. Springs; 11. Revolutions counter; 12. Speedometer; 13. Intercom No. 3 for driver; 14. Starter button; 15. Hatch cover lashing; 16. Signal button; 17. Shaft for spring of forward suspension; 18. Control lever; 19. Switch lever; 20. Ancillary fuse-box with switchboard for electrical equipment.

The armour had an advantageous all-round slope and was thick; at the hull front 90 mm with 23° slope, at the front of the superstructure 120 mm with 35° slope, at the sides 90 mm with 45° slope and at the rear 40 mm with 83° slope. At the left was the commander's flat cupola with a 12.7-mm DShK M.1938 anti-aircraft MG, left of it a ventilator below an

An SU-100 in flames during the fighting in Brandenburg at the end of April 1945. It was reported that up to the end of the fighting at the beginning of May 1945, four tank armies lost fifty-five SP guns of this type.

armoured cover. The door at the rear enabled the four crew to enter and leave the vehicle easily and to load ammunition. The magazine could hold thirty-six fixed rounds (each 30.5 kg). Together with 630 litres of fuel, the SU-101 weighed 34.1 tonnes. The SU-102, tried out in April 1945 with the 122-mm D-25-S L/48gun, weighed 34.8 tonnes. This advanced no further than experiments. After that the development of SP guns on T-34 and T-44 chassis came to an end.

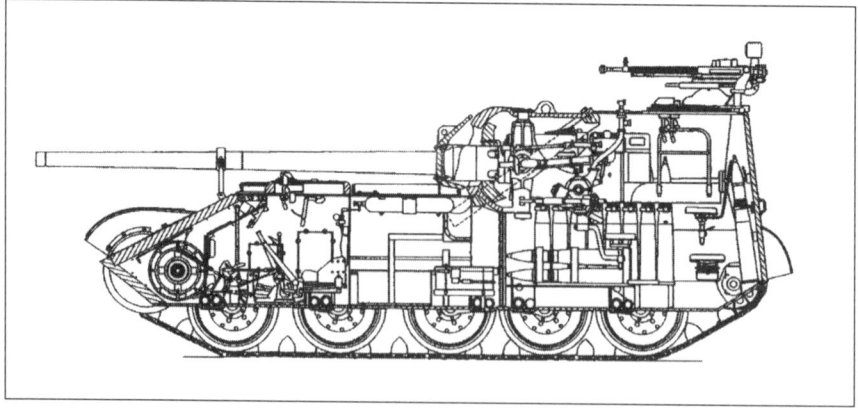

Cross-section drawing of SU-101.

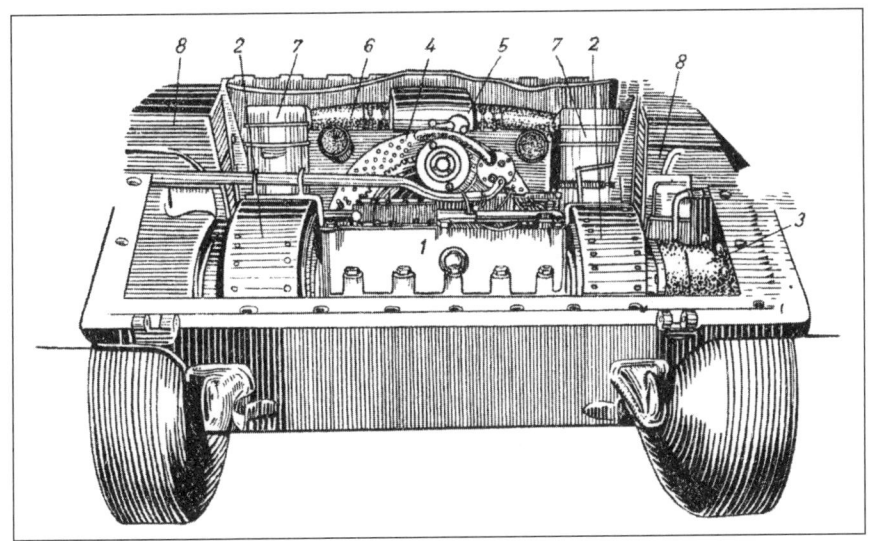

SU-100 drive compartment.
Key: 1. Transmission gear; 2. Steering clutch with brake band; 3. Lateral reduction gear; 4. Master clutch with paddle-wheel fan; 5. Air intake; 6. Piping; 7. Air filter; 8. Fuel tank.

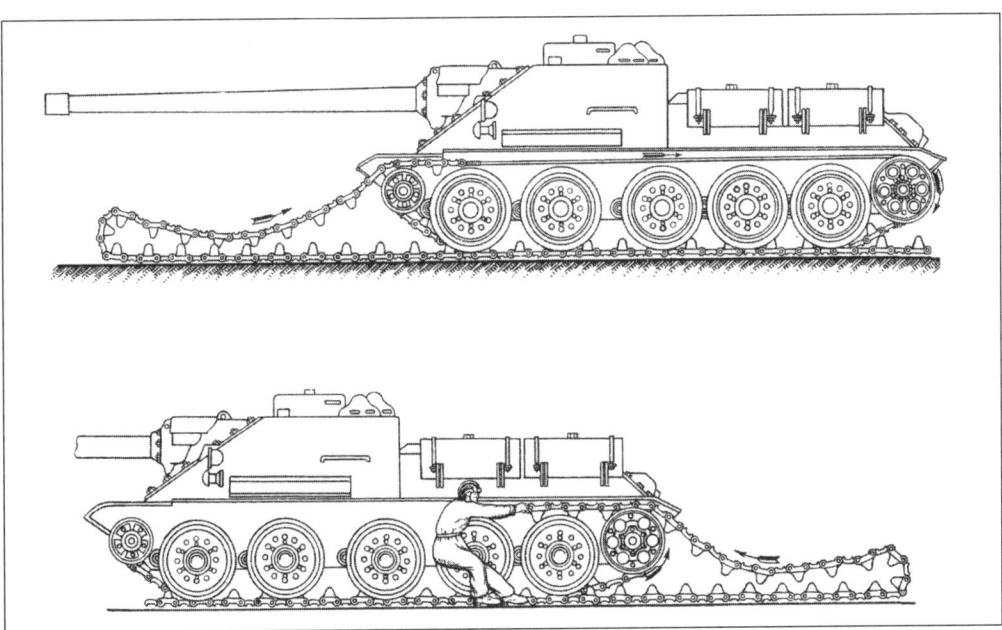

The fitting of a 1,070-kg tank track consisting of seventy-two segments using the drive wheel and a steel hawser.

T-34 Data

Details follow of the two pre-war types on which the T-34 design was based, the A-20 and A-32, plus thirteen Soviet tanks involved in the fighting against Hitler's Germany. These are made up of seven types of T-34 and one later version, the T-44, two flamethrower tanks on the T-34 chassis, three SU guns and the A-20 and A-32. All were built by various manufacturers in the Soviet Union. Unless otherwise stated all had the following standard features:

> *Engine:* Only the T-34M was not equipped with the W-2 500-hp 4-stroke diesel (60° V12) of 38.88 cm^3 capacity.
>
> *Wheels and tracks:* All tracked tanks were fitted with tracks of 72 segments, each 38.5 cm long and 50 cm broad. All had a pair of drive wheels at the rear of the track, a pair of leading or steering wheels at the front and five pairs of roller wheels in between (except for T-34M [A-43] which had six pairs of rollers and four pairs of support wheels.) The turning circle for all tanks (apart from the A-32, T-34M, A-43 and A-20, which were not reported) was 7.7 m.
>
> *Gears:* All had four forward and one reverse (except the T-34/85 which had five forward and one reverse and T-34M [A-43] which had eight forward and two reverse).
>
> *Suspension:* Independent coil springs were fitted on all models, except T-34M and T-44 which had torsion bar suspension.

A-20 Medium Tank on Wheels/Tracks

Manufacturer	Locomotive Factory Comintern, Works No. 183, Kharkov
Built	1938–9
Combat weight	18–18.2 tonnes
Engine	Standard
Length/breadth /height	5.76 m × 2.65 m × 2.435 m
Ground clearance	400–410 mm
Turning circle	n/a
Top speed	74.4 km/h (wheels), 57–65 km/h (tracks)
Range	400–700 km (wheels), 400–900 km (tracks)
Fuel tank capacity	505 litres
Fuel consumption per 100 km	n/a
Gradient/Vertical obstacle	40°/1.2 m
Fording depth	1.4 m
Specific ground pressure	0.61 kg/cm^2
Armament	1 × 45-mm M.1938 L/46 (152 rounds) 2 × 7.62-mm DT MG (2,709 rounds)
Hull armour, thickness and slope	Front: 20 mm, 33–34° Sides: 20 mm, 65° Rear: 16 mm, 45° Roof and floor: 10 mm
Turret armour, thickness and slope	Shield and front: 25 mm, curved Sides and rear: 25 mm, 75° Roof: 10 mm
Crew	4

A-32 Medium Tank

Manufacturer	Locomotive Factory 'Comintern', Works No. 183, Kharkov
Built	1939
Combat weight	19 tonnes
Engine	Standard
Length/breadth/height	5.76 m × 2.65 m × 2.435 m
Ground clearance	400 mm
Top speed	65–70 km/h
Range	420–440 km
Fuel tank capacity	460 litres
Fuel consumption per 100 km	120–180 litres
Gradient/Vertical obstacle	40°/1.20 m
Fording depth	1.4 m
Specific ground pressure	0.55 kg/cm^2
Armament	1 × 76.2-mm L-10 L/23.7 (72 rounds) 2 × 7.62-mm DT MG (1,658 rounds)
Hull armour, thickness and slope	Front: 20 mm, 33–34° Sides: 20 mm, 65° Rear: 16 mm, 45° Roof and floor: 10 mm
Armour, turret, thickness and slope	Shield: 25 mm, curved Front: 25 mm, curved Sides and rear: 25 mm, 75° Roof: 10 mm
Crew	4

T-34 Model 1940 Medium Tank	
Manufacturer	Locomotive Factory 'Comintern', Works No. 183, Kharkov
Built	1940
Combat weight	26.8 tonnes
Engine	Standard
Length/breadth/height	5.92 m × 3 m × 2.4 m
Ground clearance	400 mm
Top speed	55 km/h
Range	250–300 km
Fuel tank capacity	460 litres
Fuel consumption per 100 km	120–180 litres
Gradient/Vertical obstacle	35°/0.90 m
Fording depth	1.12 m
Specific ground pressure	0.62 kg/cm^2
Armament	1 × 76.2-mm L-11 L/30.5 (77 rounds) 2 × 7.62-mm DT MG (2,898–4,725 rounds) 1 × 7.62-mm SMG, grenades
Hull armour, thickness and slope	Front: 45 mm, 30–40° Sides: 45 mm, 90° and 40 mm, 50° Rear: 45 mm, 43–45° Roof: 16 mm
Floor	13–16 mm
Turret armour, thickness and slope	Shield and front: 45 mm, curved Sides and rear: 45 mm, 60° Roof: 15 mm, 0–5°
Crew	4

T-34M (A-34) Medium Tank	
Manufacturer	Locomotive Factory 'Comintern', Works No. 183, Kharkov
Built	1941
Combat weight	25.5 tonnes
Engine	W-5 600-hp 4-stroke diesel 12/V-form (60°), 2,000 revs/min; 38,880 cc
Length/breadth/height	5.925 mm (with gun overhang 6.43 m) × 2.75 m × 2.288 m
Ground clearance	450–500 mm
Top speed	60.5 km/h
Range	330–600 km
Fuel tank capacity	n/a
Fuel consumption per 100 km	n/a
Gradient/Vertical obstacle	35–45°/1 m
Fording depth	1.5 m
Specific ground pressure	0.60 kg/cm^2
Armament	1 × 76.2-mm F-34 L/41.5 (103 rounds) 2 × 7.62-mm DT MG (4,536 rounds) 1 × 7.62-mm SMG, grenades
Hull armour, thickness and slope	Front: 45 mm, 37–53° Sides: 45 mm, 90° Rear: 40 mm, 45–85° Roof: 16–20 mm, 2–15° Floor: 16–20 mm
Turret armour, thickness and slope	Shield: 45 mm, curved Front: 45 mm, 60° Sides: 45 mm, 65° Rear: 45 mm, 75° Roof: 20 mm, 4°
Crew	5

T-34 Model 1941 Medium Tank	
Manufacturers	Locomotive Factory Works No. 183, Kharkov; Works No. 112, Gorky; Stalingrad Tractor Factory; Kirov Works, Chelyabinsk
Built	1941
Combat weight	29.12 tonnes
Engine	Standard
Length/breadth/height	5.9 m (with gun overhang 6.68 m) × 3 m × 2.45 m
Ground clearance	400 mm
Top speed	55 km/h
Range	250–465 km
Fuel tank capacity	460 litres
Fuel consumption per 100 km	120–180 litres
Gradient/Vertical obstacle	35°/0.9 m
Fording depth	1.3 m
Specific ground pressure	0.72 kg/cm^2
Armament	1 × 76.2-mm F-34 L/41.5 (77 rounds) 2 × 7.62-mm DT MG (2,646–2,394 rounds) 1 × 7.62-mm SMG, grenades.
Hull armour, thickness and slope	Front: 45 mm, 30–37° (driver's hatch, 72 mm) Sides: 40 mm, 50° and 45 mm, 90° Rear: 40 mm, 42–45° Roof: 16 mm Floor: 13–16 mm, later 16–20 mm
Turret armour, thickness and slope	Shield: 45 mm, curved Front: 45 mm, later 52 mm, curved Sides: 45 mm, later 52 mm, 60° Rear: 45 mm, later 52 mm, 60° Roof: 15 mm
Crew	4

T-34 Model 1942 Medium Tank	
Manufacturer	Works No. 183, Nizhny Tagil; Works No. 112, Gorky; Works No. 174, Omsk; Uralmash, Sverdlovsk; Kirov Works, Chelyabinsk; Stalingrad Tractor Factory
Built	1942–3
Combat weight	28.5–29.1 tonnes
Engine	Standard
Length/breadth/height	5.92 m (with gun overhang 6.68 m) × 3 m × 2.45 m
Ground clearance	400 mm
Top speed	55 km/h
Range	250–465 km
Fuel tank capacity	610 litres
Fuel consumption per 100 km	120–180 litres
Gradient/Vertical obstacle	35°/0.9 m
Fording depth	1.3 m
Specific ground pressure	0.72 kg/cm^2
Armament	1 × 76.2-mm F-34 L/41.5 (78–101 rounds) 2 × 7.62-mm DT MG (2,235–3,150 rounds) 1 × 7.62-mm SMG, grenades
Hull armour, thickness and slope	Front: 45 mm, 30°–37° (driver's hatch 72 mm and in places 92 mm) Sides: 40 mm, 50° and 45 mm, 90° Rear: 40 mm, 42–45° Roof: 16–20 mm Floor: 16–20 mm
Turret armour, thickness and slope	Shield: 45 mm, curved Front: 52 mm, 60° Sides: 52 mm, 70° Rear: 52 mm, 70° Roof: 20 mm
Crew	4

T-34 Model 1943 Medium Tank	
Manufacturer	Works No. 112, Gorky; Works No. 174, Omsk; Works No. 183, Nizhny Tagil
Built	1943–4
Combat weight	30.5–30.9 tonnes
Engine	Standard
Length/breadth/height	6.1 m (with gun overhang 6.75 m) × 3 m × 2.52 mm
Ground clearance	400 mm
Top speed	53.6–55 km/h
Range	400–465 km
Fuel tank capacity	610 litres
Fuel consumption per 100 km	120–180 litres
Gradient/Vertical obstacle	35°/0.73 m
Fording depth	1.3 m
Specific ground pressure	0.72 kg/cm^2
Armament	1 × 76.2-mm F-34 L/41.5 (77–100 rounds) 2 × 7.62-mm DT MG (2,275–3,000 rounds) 1 × 7.62-mm SMG, grenades
Hull armour, thickness and slope	Front: 45 mm, 30–37° (driver's hatch 72 mm) Sides: 40 mm, 50° and 45 mm, 90° Rear: 40 mm, 42–45° Roof: 20 mm Floor: 16–20 mm
Turret armour, thickness and slope	Shield: 45 mm, curved Front: 52 mm, 60° Sides: 52 mm, 70° Rear: 52 mm, 70° Roof: 20 mm
Crew	4

T-34/85 Model 1943 Medium Tank	
Manufacturer	Works No. 112, Gorky; Works No. 174, Omsk; Works No. 183, Nizhny Tagil
Built	1943–4
Combat weight	32.7 tonnes
Engine	Standard
Length/breadth/height	6.19 m (with gun overhang 8.11 m) × 3 m × 2.743 m
Ground clearance	400 mm
Top speed	55 km/h
Range	350–420 km
Fuel tank capacity	755 litres
Fuel consumption per 100 km	180–220 litres
Gradient/Vertical obstacle	30°/0.73 m
Fording depth	1.3 m
Specific ground pressure	0.84 kg/cm^2
Armament	1 × 85-mm D5T L/51.6 (55–60 rounds) 2 × 7.62-mm DT MG (1,953–2,746 rounds) 1 × 7.62-mm SMG, grenades
Hull armour, thickness and slope	Front: 45 mm, 30° Sides: 45 mm, 60–90° Rear: 45 mm, 42–45° Roof: 20 mm Floor: 20 mm
Turret armour, thickness and slope	Shield: 90 mm, curved Front: 90 mm, curved Sides: 75 mm, 70° Rear: 52 mm, 80° Roof: 20 mm
Crew	5

T-34/85 Model 1944 Medium Tank	
Manufacturer	Works No. 112, Gorky; Works No. 174, Omsk; Works No. 183, Nizhny Tagil
Built	1944–6
Combat weight	32.12–32.23 tonnes
Engine	Standard
Length/breadth/height	6.19 m (with gun overhang 8.115 m) × 3 m × 2.743 m
Ground clearance	400 mm
Top speed	52.6–54.4 km/h
Range	290–350 km
Fuel tank capacity	650 litres
Fuel consumption per 100 km	460 litres
Gradient/Vertical obstacle	30°/0.7 m
Fording depth	1.3 m
Specific ground pressure	0.84–0.86 kg/cm^2
Armament	1 × 85-mm ZIS-S-53 L/54.6 (56–60 rounds) 2 × 7.62-mm DT MG (1,953–2,079 rounds) 1 × 7.62-mm SMG, grenades
Hull armour, thickness and slope	Front: 45–60 mm, 30° Sides: 45 mm, 60–90° Rear: 45 mm, 42–45° Roof: 20 mm Floor: 20 mm
Turret armour, thickness and slope	Shield: 90 mm, curved Front: 90 mm, curved Sides: 75 mm, 70° Rear: 52 mm, 80° Roof: 20 mm
Crew	5

T-44 Medium Tank	
Manufacturer	Works No. 183 and Works No. 75, Kharkov
Built	1944–6
Combat weight	30.4 tonnes
Engine	Standard
Length/breadth/height	6 m (with gun overhang 7.65 m) × 3.1 m × 2.47 m
Ground clearance	430–455 mm
Top speed	52.6 km/h
Range	210–235 km
Fuel tank capacity	650 litres
Fuel consumption per 100 km	460 litres
Gradient/Vertical obstacle	30°/0.7 m
Fording depth	1.3 m
Specific ground pressure	0.85 kg/cm²
Armament	1 × 85-mm ZIS-S-53 L/54.6 (51–58 rounds) 2 × 7.62-mm DT MG (1,890–1,953 rounds) 1 × 7.62-mm SMG, grenades
Hull armour, thickness and slope	Front: 75 mm, 30–45° Sides: 75 mm, 90° Rear: 30–45 mm, 20–90° Roof: 15 mm Floor: 15 mm
Turret armour, thickness and slope	Shield: 115 mm, curved Front: 115 mm, curved Sides: 90 mm, 70° Rear: 75 mm, 78° Roof: 15 mm
Crew	4

OT-34 Flamethrower Tank	
Manufacturer	Works No. 183, Nizhny Tagil; Works No. 112, Gorky; Works No. 174, Omsk, from 1943
Built	1942–4
Combat weight	31 tonnes
Engine	Standard
Length/breadth/height	6.1 m (with gun overhang 6.75 m) × 3 m × 2.4 m
Ground clearance	400 mm
Top speed	55 km/h
Range	n/a
Fuel tank capacity	n/a
Fuel consumption per 100 km	120–180 litres
Gradient/Vertical obstacle	35°/0.73 m
Fording depth	1.3 m
Specific ground pressure	0.84 kg/cm^2
Armament	1 × 76.2-mm F-34 L/41.5 (71–77 rounds) 1 × 7.62-mm DT MG (2,394–2,646 rounds) 1 × ATO-41 flamethrower (fuel 105 litres) 1 × 7.62-mm SMG, grenades.
Hull armour, thickness and slope	Front: 45 mm, 30–37° Sides: 45 mm, 90°, and 40 mm, 50° Rear: 42–45 mm Roof: 20 mm Floor: 16–20 mm
Turret armour, thickness and slope	Shield: 45 mm, curved Front: 52 mm, 60° Sides: 52 mm, 80° Rear: 52 mm, 80° Roof: 20 mm
Crew	3

OT-34/85 Flamethrower Tank	
Manufacturer	Works No. 183, Nizhny Tagil
Built	1944–5
Combat weight	32 tonnes
Engine	Standard
Length/breadth/height	6.1 m (with gun overhang 8.1 m) × 3 m × 2.7 m
Ground clearance	400 mm
Top speed	53.6–55 km/h
Range	300 km
Fuel tank capacity	n/a
Fuel consumption per 100 km	120–180 litres
Gradient/Vertical obstacle	35°/0.73 m
Fording depth	1.3 m
Specific ground pressure	0.85 kg/cm^2
Armament	1 × 85-mm ZIS-S-53 L/54.6 (58 rounds)
	1 × 7.62-mm DT MG (1,827 rounds)
	ATO-42 flamethrower (fuel 200 litres)
	1 × 7.62-mm SMG, grenades
Hull armour, thickness and slope	Front: 45–60 mm, 30°
	Sides: 45 mm, 60–90°
	Rear: 45 mm, 42–45°
	Roof: 20 mm
	Floor: 20 mm
Turret armour, thickness and slope	Shield: 90 mm, curved
	Front: 90 mm, curved
	Sides: 75 mm, 70°
	Rear: 52 mm, 80°
	Roof: 20 mm
Crew	4

SU-122 Self-Propelled Gun	
Manufacturer	Uralmash Works, Sverdlovsk
Built	1944–6
Combat weight	29.6 tonnes
Engine	Standard
Length/breadth/height	6.1 m (to muzzle 6.95 m) × 3 m × 2.235 m
Ground clearance	400 mm
Top speed	55 km/h
Range	260–600 km
Fuel tank capacity	860 litres
Fuel consumption per 100 km	165–270 litres
Gradient/Vertical obstacle	33°/0.73 m
Fording depth	1.3 m
Specific ground pressure	0.68–0,72 kg/cm^2
Armament	1 × 122-mm M-30 Model 1938 L/22.7 howitzer (40 shells/cartridges) 2 × 7.62-mm SMG, hand and anti-tank grenades
Hull armour, thickness and slope	Front: 45–75 mm, 35–40° Sides: 45 mm, 90° Rear: 40 mm, 42–45° Roof: 20 mm Floor: 20 mm
Armour, casemate, thickness and slope	Shield: 75 mm, curved Front: 75 mm, 40° Side: 45 mm, 70° Rear: 45 mm, 90° Roof: 20 mm
Crew	5

SU-85 Self-Propelled Gun	
Manufacturer	Uralmash Works, Sverdlovsk
Years when built	1943–4
Combat weight	31 tonnes
Engine	Standard
Length//breadth/height	6.1 m (to muzzle 8.13 m) × 3 m × 2.235 m
Ground clearance	400 mm
Top speed	47 km/h
Range	600 km
Fuel tank capacity	860 litres
Fuel consumption per 100 km	165–270 litres
Gradient/Vertical obstacle	33°/0.73 m
Fording depth	1.3 m
Specific ground pressure	0.73 kg/cm^2
Armament	1 × 85-mm D-5-S 85A L/51.6 (48 rounds)
	2 × 7.62-mm SMG, hand- and anti-tank grenades
Hull armour, thickness and slope	Front: 45 mm, 40–45°
	Sides: 45 mm, 90°
	Rear: 40 mm, 42–45°
	Roof: 20 mm
	Floor: 15 mm
Casemate armour, thickness and slope	Shield: 45–52 mm, curved
	Front: 45 mm, 40°
	Side: 45 mm, 70°
	Rear: 45 mm, 80°
	Roof: 20 mm
Crew	4

SU-100 Self-Propelled Gun	
Manufacturer	Uralmash Works, Sverdlovsk
Years when built	1944–6
Combat weight	31.6 tonnes
Engine	Standard
Length/breadth/height	5.93 m (to muzzle 9.45 m) × 3 m × 2.245 m
Ground clearance	400 mm
Top speed	50 km/h
Range	180–310 km
Fuel tank capacity	860 litres
Fuel consumption per 100 km	360–400 litres
Gradient/Vertical obstacle	35°/0.73 m
Fording depth	1.3 m
Specific ground pressure	0.8–0.82 kg/cm^2
Armament	1 × 100-mm D-10-S Model 1944 L/56 gun (33 rounds) 2 × 7.62-mm SMG, hand- and anti-tank grenades
Hull armour, thickness and slope	Front: 45 mm, 40–45° Sides: 45 mm, 90° Rear: 40 mm, 42–45° Roof: 20 mm Floor: 15 mm
Casemate armour, thickness and slope	Shield: 45–52 mm, curved Front: 45 mm, 50° Side: 45 mm, 70° Rear: 45 mm, 80° Roof: 20 mm
Crew	4

Also available from Greenhill Books:

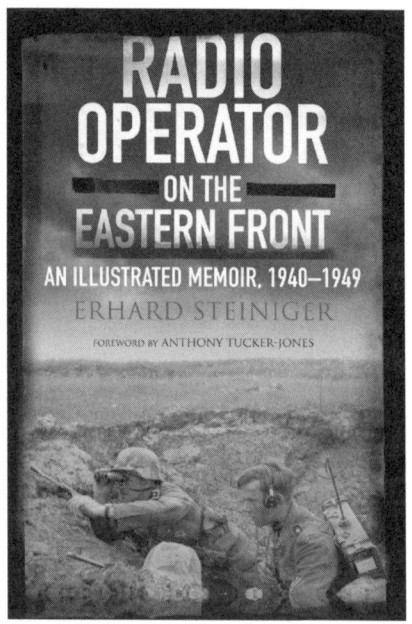

www.greenhillbooks.com